Lecture Notes in Physics

Edited by H. Araki, Kyoto, J. Ehlers, München, K. Hepp, Zürich
R. Kippenhahn, München, D. Ruelle, Bures-sur-Yvette
H. A. Weidenmüller, Heidelberg, J. Wess, Karlsruhe and J. Zittartz, Köln
Managing Editor: W. Beiglböck

341

E. F. Milone (Ed.)

Infrared Extinction and Standardization

Proceedings of two Sessions of IAU Commissions 25 and 9
Held at Baltimore, Maryland, USA, August 4, 1988

Springer-Verlag
Berlin Heidelberg New York London Paris Tokyo Hong Kong

Editor

E. F. Milone
Department of Physics and Astronomy, The University of Calgary
2500 University Drive N.W., Calgary, Alberta, Canada T2N 1N4

ISBN 3-540-51610-7 Springer-Verlag Berlin Heidelberg New York
ISBN 0-387-51610-7 Springer-Verlag New York Berlin Heidelberg

This work is subject to copyright. All rights are reserved, whether the whole or part of the material is concerned, specifically the rights of translation, reprinting, re-use of illustrations, recitation, broadcasting, reproduction on microfilms or in other ways, and storage in data banks. Duplication of this publication or parts thereof is only permitted under the provisions of the German Copyright Law of September 9, 1965, in its version of June 24, 1985, and a copyright fee must always be paid. Violations fall under the prosecution act of the German Copyright Law.

© Springer-Verlag Berlin Heidelberg 1989
Printed in Germany

Printing: Druckhaus Beltz, Hemsbach/Bergstr.
Bookbinding: J. Schäffer GmbH & Co. KG., Grünstadt
2158/3140-543210 – Printed on acid-free paper

TABLE OF CONTENTS

Problems of Infrared Extinction
and Standardization - An Introduction
E. F. Milone 1

Extinction and Transformation
A. T. Young 6

Models of Infrared Atmospheric Extinction
K. Volk, T. A. Clark, and E. F. Milone 15

Atmospheric Extinction in the Infrared
R. J. Angione 25

Infrared Extinction at Sutherland
I. S. Glass and B. S. Carter 37

Near-Infrared Extinction Measurements
at the Indian Observatory Sites
N. M. Ashok 49

Reducing Photometry by Computing Atmospheric Transmission
R. L. Kurucz 55

JHKLM Photometry: Standard Systems, Passbands
and Intrinsic Colors
M. S. Bessell and J. M. Brett 61

Standardization with Infrared Array Photometers
I. S. McLean 66

A Summary of the Session
R. A. Bell 72

Concluding Postscript
E. F. Milone 77

PROBLEMS OF INFRARED EXTINCTION AND STANDARDIZATION
- AN INTRODUCTION[1]

E. F. Milone
The Astrophysical Observatory
The University of Calgary
Calgary, AB T2N 1N4/Canada

I. Preamble

The following paragraphs contain introductory remarks to the joint meeting of Commissions 25 and 9 held during sessions 1 and 2 at the General Assembly of the International Astronomical Union on August 4, 1988 in Baltimore, Maryland. The purpose of the meeting was to explore the problems associated with atmospheric extinction and standardization of infrared flux from astronomical sources.

There has long been need for such a meeting. Although the effects of the earth's atmosphere on visual starlight have been extensively discussed, this has not been true for the atmospheric effects on the infrared radiation from astronomical sources. Moreover, the process of transforming data acquired in local systems into a standard system is more complicated than it is in the visual. The reasons for the complications and for possible remedies, both for extinction and for standardization, were explored in the meetings' papers which are reproduced in this volume.

This meeting had its genesis in a letter from myself to Prof. F. Rufener, President of Commission 25, decrying the current state of infrared data reduction and proposing a joint commission meeting to discuss it. Prof. Rufener enthusiastically endorsed the topic. He was joined by Dr. C. Humphries, President of Commission 9 and by a number of members of both commissions, a number of whom agreed to become members of the Scientific Organizing Committee. Among the supporters was Prof. A. J. Wesselink, for whose guidance since my graduate student days at Yale I continue to be grateful. The SOC which helped to organize this meeting consisted of Drs. R. A. Bell, M. S. Bessell, I. S. Glass, R. L. Kurucz, T. A. Clark, I. S. McLean, F. Rufener (ex officio), and A. T. Young. To all the speakers, especially to Dr. Young and Dr. Bell, who kindly agreed to provide the major initial and concluding addresses, respectively, I express thanks for both the excellent quality of presentations and the splendid cooperation which made possible the meetings and this volume.

---

[1] With Postscript, publication of the Rothney Astrophysical Observatory, No. 55.

II. Opening Remarks

In considering the problems of infrared extinction and standardization, we may be facing cases of opposing perceptions. In the dialectic of life, if not history, it is often the case that opposing viewpoints as well as models of reality, are in fact reconcilable into new and powerful syntheses. At least for a short while, such syntheses may be able to capture the spirit, or at least command the respect, of the communities whose activities they purport to model. The present task is to bring to bear sufficient light on these problems so that they can be recognized and possible directions for finding their solutions can be identified and agreed upon.

First, given the available atmospheric windows, sketched in Figure 1 (reproduced from Clark and Irwin 1973) the range 1-30 μm should suffice as a working definition of the ground-based infrared.

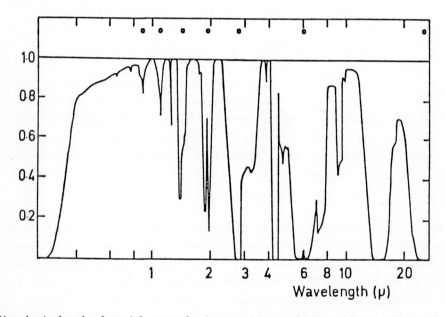

Fig. 1 A sketch of zenith atmospheric transmission between the visible region and the infrared window at 20 μm, at an altitude of 2 km and precipitable water vapor content of 0.25 cm. The 20 μm window becomes effectively opaque for a water vapor content of 0.5 cm. The small circles indicate major water vapor absorption bands. Reproduced from Clark and Irwin (1973) with permission.

In the absence of preordained guidelines, most of the speakers assumed narrower wavelength limits, reflecting, presumably, the chief interests of the community. Indeed, most of the papers concentrate on the JHKL region. Moreover, none of our papers deal with balloon-, rocket- or space-infrared standardization problems directly. The transforming of space observations into ground-based systems is in itself difficult, and its accomplishment certainly deserves extensive discussion in other forums. Despite these omissions, what has been done here is a significant and useful step forward on the road to a better understanding of both the problems and the solutions of infrared extinction and standardization.

III. Focal Points

I have been impressed, since the early 1970's with the intrinsic quality of infrared photometry, specifically with the ability of infrared photometers to produce excellent internal precision over short intervals of time. On occasion, the sky is so temporally and spatially uniform that the Bouguer method can yield apparently trustworthy results even to large values of the air mass. Figures 2 and 3 depict InSb K observations of the star Regulus obtained at KPNO on the 1.3-m telescope in June 1972.

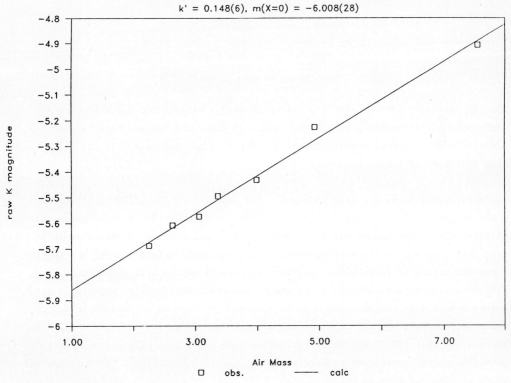

Fig. 2 Extinction observations of α Leo in the K-band from KPNO. The air mass includes a correction for atmospheric curvature but it has not been tailored to the appropriate scale height of water vapor.

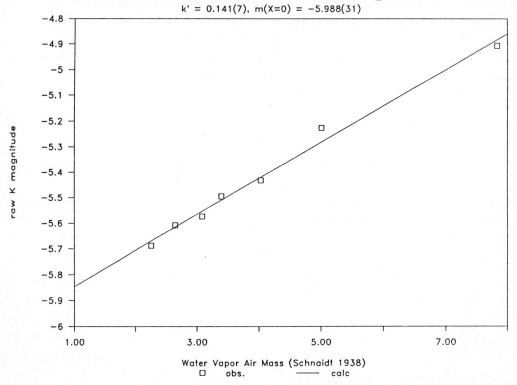

Fig. 3 The same observations of α Leo as in Fig. 2 but plotted against relative water vapor mass (Schnaidt 1938), the values of which are intermediate between Bemporad's (1904) tabular data and sec z. For z < 74°, (sec z < 3.7) the differences from the Bemporad tables are < 1%.

The data extend to an air mass of about 8 (at which point the effect of different scale heights of the extinction agents is very important). The work of which this illustration is an example yielded linear extinction coefficients with apparently good precision and led to what appeared to be reasonable outside-atmosphere magnitudes and colors (Milone 1976). However, as several papers will demonstrate, such determinations do not tell the whole story, and while differential magnitudes and colors may be determinable from sources observed at the same air masses in an apparently uniform sky on the same night, the Bouguer zero-points may not be. The Manduca and Bell (1979) synthetic extinction work explicitly raised some unsettling questions about our ability to obtain such values from traditional (from an optical point of view) sources of data.

My own attempt to deal with this problem was to attempt to predict the Manduca and Bell zero-air mass magnitudes by carefully fitting the synthetic data between 1 and

3 air masses with combinations of polynomial, square-root and log-term functions. I was moderately successful (i.e. to the 2-3% level) in doing this, presumably because of subtle effects of the low-air mass portion of the curve on the 1-3 air mass domain. But fitting real, and therefore even slightly noisy data has not been successful, and, as we shall see, there are physical reasons to explain why this technique should not work. Therefore, some questions have to be asked. What data do we need in order to have a practical way of doing absolute photometry to the 1% or better level, or even differential photometry when sources are at disparate air masses? Is it sufficient to observe one or more standards at similar air masses to the program star and, regardless of extinction variations from night to night, expect thereby, to be able to transform and to combine such observations? This is common practice; at some level it seems to work, and indeed, one may ask what else one may do. This raises two other questions. What are the consequences of this practice? Can other options be found?

As still other speakers will argue, standardization is a major problem in the infrared as it is in other wavelength regions of broad-band photometry. Has this problem been compounded in the infrared because of the treatment of extinction? Will it help for the community to agree on calibrating standards and sources, and filters, and if so, whose? Can we obtain agreement to publish the details of our photometer systems as an aid to others in transforming data from one system to another?

We have a stimulating array of presentations expressing wide-ranging opinions on these and other relevant questions. I know of no one better suited to issue the challenge to complacency than Andy Young. He not only spells out the problem of extinction, he also suggests a solution. His presentation is followed by papers of both theoreticians and observers and include reports of work on absorption band and spectral synthesis, a progress report on his work by Bob Kurucz, and a discussion on standardization with infrared array detectors by Ian McLean. The papers have been arranged in two broad sections: extinction and standardization; in each section the mix of presentations provides a hopefully balanced view of both how infrared photometry is done and perhaps how it may be improved. The excellent summary is provided by Roger Bell, who, despite his disclaimer, is an important and well-known contributor to the field.

References

Bemporad, A., 1904. Mitt. d. Sternwarte Heidelberg, No. 4.
Clark, T. A., and Irwin, G., 1973. J. Roy. Astron. Soc. Canada, 67, 142.
Manduca, A., and Bell, R. A., 1979. Publs. Astron. Soc. Pac., 91, 848.
Milone, E. F., 1976. Astrophys. J. Suppl., 31, 93.
Schnaidt, F., 1938. Meteorologische Zeitschrift, 55, 296.

EXTINCTION AND TRANSFORMATION

Andrew T. Young
Astronomy Department, San Diego State University
San Diego, California 92182-0334

Abstract

The basic principles of heterochromatic extinction show that the approach used in the visible should not work well in the infrared, where molecular line absorption rather than continuous scattering dominates the extinction. Not only does this extinction change very rapidly with wavelength (so that stellar color becomes only weakly correlated with effective extinction), but also many of the lines are saturated (so that Forbes's curve-of-growth effect is much more severe in the IR.) Furthermore, broadband IR colors are more undersampled than those in the visible, so aliasing errors make them correlate even less with extinction, and enhance the difficulties of transformation to a standard system. Reduction to outside the atmosphere is difficult, but a rational approximation for the Forbes effect may help. Plausible assumptions about the probability distribution function of line strengths, and band-model approaches, may be useful. The only solution to the transformation problem is to satisfy the sampling theorem, which may be difficult in the IR because of gaps due to saturated telluric absorptions.

Introduction

Correction for heterochromatic extinction is an incompletely solved problem of long standing. When the monochromatic extinction coefficient varies within the instrumental passband, the Bouguer curve is nonlinear, as was first noticed by Forbes (1842). We may regard the Forbes effect as a curve-of-growth effect: the extinction first becomes saturated at the wavelengths where it is largest, leaving only those with weaker extinction to affect the Bouguer curve at larger air masses. Thus, the reduction in slope of the Bouguer curve at large air masses is analogous to the reduction in slope of the curve of growth as an absorption line saturates, although the quantities plotted are quite different in the two cases. Mathematically, the same situation is encountered in nuclear physics and fluorescence spectroscopy, where two or more different decay times are present. In our case, air mass plays the role of time; we may think of the number of photons transmitted at each wavelength as decaying exponentially with increasing air mass.

As the wavelengths with large extinction coefficients are removed first, the transmitted intensity becomes dominated by the wavelengths with the least monochromatic extinction. At large air masses, the slope of the Bouguer curve approaches the smallest monochromatic extinction coefficient within the instrumental passband as a limit. This means that the slope of the curve changes most rapidly between airmasses of 0 and 1. Thus, in practice, it is often extremely difficult to distinguish the observed Bouguer curve from a straight line, even though its linear extrapolation to zero air mass may be quite removed from the true value. This has been noticed in the U band of the UBV system (Hardie, 1966; Young, 1974, p.159) as well as in infrared bands like JHKL (Manduca and Bell, 1979).

Why what works in the visible doesn't work in the IR

I know of no general theoretical treatment for this problem. In the visible, extinction is mainly due to scattering by molecules and aerosols. These processes are smooth functions of wavelength, so they can be represented fairly well by Taylor expansions about a centroid wavelength (Strömgren, 1937; King, 1952; Young, 1974, 1988). Furthermore, because the scattering varies monotonically with wavelength, the effective extinction coefficients for different stars depend linearly on the stars' spectral gradients, which are fairly well approximated by color indices.

This "color term" in the extinction comes mathematically from the second-order term in the Taylor expansion of the transmitted spectral irradiance $S(\lambda) = I(\lambda) \times t(\lambda)$, where I is the extra-atmospheric stellar spectral irradiance and t is the atmospheric transmission. (The first-order term vanishes because the expansion is about the centroid of the instrumental response function.) We have

$$S'' = I''t + I't' + It'' \tag{1}$$

where the primes denote wavelength derivatives. This expression appears in the integral over wavelength of the received spectral irradiance, weighted by the instrumental response function R. Thus, the term it contributes to the measured quantity is essentially the cross correlation of the wavelength gradients of the stellar irradiance and the atmospheric extinction.

As wavelength is an alias for both the monochromatic stellar intensity and the monochromatic extinction coefficient, the integration over wavelength effectively gives the correlation between stellar spectral irradiance and monochromatic extinction. Hence, we can use stellar irradiance gradients (approximated by color indices) to compute both the variation of extinction coefficient from star to star (the dependence of extinction on spectral type, first noticed by Guthnick and Prager in 1914) as well as the variation of extinction coefficient with atmospheric reddening, and hence airmass (the Forbes effect).

However, in the infrared, both Rayleigh and aerosol scattering are very small, if not negligible, and the extinction is dominated by molecular absorption. Unfortunately, this has a great deal of fine structure, due to the rotation-vibration lines of individual molecular bands. In this case, the monochromatic extinction coefficient within the instrumental passband is only weakly correlated with wavelength, so stars of different colors have rather similar extinction coefficients at a given airmass, even though the extinction coefficient varies strongly with airmass (i.e., the Forbes effect is very large).

If there is no correlation of extinction with wavelength, we cannot obtain the information we need from stellar colors. This seems to be a very good assumption for bands like H, K, and L, though it is rather poor for J. Let us suppose this is so, and ask what physical variables determine the problem, and how we may represent them.

Mathematical representation

Very crudely, let us assume a rectangular instrumental response function, which includes both a constant background extinction (due to aerosols, the water-vapor-dimer continuum, and the like), and a molecular band, represented as rectangular functions of wavelength. Then there are three atmospheric variables: the fraction of the instrumental bandpass occupied by the molecular band; the extinction coefficient inside the band; and the extinction coefficient outside the band.

The Bouguer curve then requires these three parameters to describe its shape, and a fourth, the star's magnitude outside the atmosphere, to place it vertically on the magnitude scale. Thus, we must solve for a minimum of four parameters. From the photometrist's point of view, these are the height, the mean slope, the mean curvature, and the *change* of curvature with increasing airmass.

A more realistic model would represent the (response-weighted) probability density distribution of absorption coefficient within the instrumental passband more accurately, and would probably require still more parameters; band models of molecular absorption might be useful. I cannot see how any simpler picture could be adequate.

This is a dismal prospect, as observers in the visible have enough trouble to determine the two parameters that describe their linear Bouguer plots. This seemingly intractable problem has led me to avoid infrared work for the past two decades. However, it now appears a bit less intractable than before. To begin with, all the numerically calculated extinction curves show a remarkable qualitative similarity: a steep slope at zero airmass, followed by an asymptotic approach to a constant slope at large airmasses. This suggests approximating the extinction curve by the sum of its linear asymptote and some rapidly-decaying function of the airmass, M. If (as is usually the case) only a small part of the light is affected by the molecular absorption, we might take this latter function to be a negative exponential.

Trial fits of Manduca and Bell's (1979) numerical values to

$$\Delta m = aM + b + c\exp(-dM) \qquad (2)$$

gave fairly good fits, with typical residuals of one or two millimagnitudes, and largest residuals of 3 or 4 millimagnitudes. However, an even better function for the decaying part is $\frac{1}{M + M_0}$, which allows the whole curve to be represented by the more general function (a Padé approximant)

$$\Delta m = \frac{(AM^2 + BM + C)}{(M + M_0)} \qquad (3)$$

from $M = 0$ to 3, with maximum residuals less than 1.8 millimag.

Table I. Extrapolation errors (mag) at M = 0 from fitting Eq. (3) to Manduca and Bell's values from M = 1.0 to 3.0

	Kitt Peak filters		Johnson filters		
Δ (Vega)	J	K	J	K	L
Kitt Peak, summer	+0.016	+0.003	-0.037	+0.051	+0.010
Kitt Peak, winter	+0.038	+0.003	+0.037	+0.017	+0.027
Mauna Kea	-0.007	-0.059	+0.026	+0.012	
Δ (cool giant)					
Kitt Peak, summer	+0.058	-0.000	+0.081	+0.001	
Kitt Peak, winter	-0.022	+0.006	+0.071	+0.027	
Mauna Kea	+0.026	-0.000	+0.034	+0.029	
δ (giant - Vega)					
Kitt Peak, summer	+0.042	-0.003	+0.118	-0.050	
Kitt Peak, winter	-0.060	+0.003	+0.040	+0.010	
Mauna Kea	+0.033	+0.059	+0.008	+0.017	

A consistent trend in the residuals shows that this form, though good, does not represent the artificial data perfectly: the residuals at $M = 0$, 1.0, and 3.0 are almost always positive; and the one at $M = 1.0$ is nearly always the largest. This causes systematic errors when one tries to fit Eq. (3) to just the inside-the-atmosphere data. In particular, the value of M_0 estimated from the observable range of airmasses is usually too large, which makes the estimated intercept at $M = 0$ too large (faint) as well.

One might at first be encouraged by the errors in the extrapolated points at $M = 0$, which are typically 3 or 4 times smaller than those (called Δ in Manduca and Bell's (1979) Table III) produced by linear extrapolation from the points at $M = 1$ and 2 (see Table I). But the differences between the extrapolations from the Vega and giant models (called δ by Manduca and Bell) are about twice as large as those from the linear extrapolation. The reason is that the values extrapolated from Eq. (3) are so sensitive to the data that even rounding the theoretical values to the nearest millimagnitude produces extrapolation errors of several hundredths of a magnitude. Mathematically, the problem is so ill-posed that the noise in the data is magnified nearly a hundredfold in the intercept, and more than a thousandfold in the value of M_0.

Table II. True values of M_0 found by fitting Eq. (3) to Manduca and Bell's values from M = 0.0 to 3.0 (upper row of each pair) compared to those found by fitting only the values from M = 1.0 to 3.0 (lower row of each pair)

Vega model	Kitt Peak		Johnson filters		
	J	K	J	K	L
Kitt Peak, summer	0.522	1.829	0.521	0.832	1.405
	0.737	2.589	0.500	1.862	1.708
Kitt Peak, winter	0.730	2.586	0.724	1.294	1.897
	2.485	4.057	1.478	2.105	4.375
Mauna Kea	0.916	3.227	0.879	1.548	
	0.851	0.279	1.863	2.598	
cool giant model					
Kitt Peak, summer	0.468	1.786	0.479	0.803	
	1.023	2.257	0.957	0.957	
Kitt Peak, winter	0.687	2.305	0.693	1.300	
	0.655	4.659	2.411	2.513	
Mauna Kea	0.855	2.397	0.868	1.518	
	2.282	1.694	1.830	6.674	

However, this very fact means that we can use an *assumed* value of M_0 and not do much violence to the modelling. In turn, this means that we are only determining three parameters from observational data instead of four, so there may be some hope of getting useful results in actual practice. The "true" values of M_0 from the full fits, including all the artificial data out to $M = 0$, ranged only from 0.47 to 0.92 for the J filters, from 0.80 to 1.55 for the Johnson K filter, and from 1.79 to 3.23 for the Kitt Peak K filter (see Table II). Furthermore, they progressed smoothly toward larger values as the molecular absorption diminished, increasing in J about 0.2 from summer (wet) to winter (dry) conditions at Kitt Peak, and again by another 0.2 from there to Mauna Kea. The change from Vega to the cool giant model was much smaller, about 0.06. Changing from the Kitt Peak to the Johnson J filters produced even less change in the "true" value of M_0.

Thus, reasonable guesses for M_0 at Kitt Peak are about 0.6 in the J band; 1.0 for Johnson K; and 1.6 for Johnson L. At Mauna Kea, values about 50% larger are appropriate. Table III shows the extrapolation errors Δ and the differences between the giant and Vega, δ, assuming such values for M_0, and fitting Eq. (3) to just the $M = 1.0$ to 3.0 data. The error in the guessed value of M_0 propagates into the intercept error, but reduced by about an order of magnitude.

The largest Δ, assuming a reasonable M_0, is 0.033 mag, and the absolutely largest difference between giant and Vega models δ is 0.015 mag. These errors are comparable to those often regarded as acceptable in UBV photometry. Even if we ignored the model calculations entirely and assumed $M_0 = 1.0$ in every case, the largest errors in Δ are 0.086 mag., and the largest δ is 0.030 (both for Johnson J at Kitt Peak in the summer, for the giant model). These values are still about 3 times smaller than the corresponding ones for the linear fits given in Table III of Manduca and Bell (1979). Thus, using Eq. (3) with even crude guesses for M_0 should allow substantially better extinction corrections than the usual assumption of a linear Bouguer curve.

Because the large-airmass asymptotic slope of the extinction curve, a/M_0, is (on the simple argument given above) the minimum monochromatic extinction within the instrumental passband, it should be the same for all stars. This is very nearly true in the fits to Manduca and Bell's artificial data, even in the J band. However, the slope at $M = 0$ (i.e., $\frac{b-(c/M_0)}{M_0} \approx \frac{b}{M_0}$ because $c \approx 0$ for these data) is much larger for the giant than for Vega. Thus, a "color term" in b is necessary for the J band; possibly an additional one in a would be useful.

Transformations and sampling

As King (1952) pointed out, such color terms in extinction corrections really represent a color transformation between the inside- and outside-the-atmosphere systems. However, as I have recently shown (Young, 1988), this transformation can be done accurately only if good approximations are used for the derivatives that are involved. Undersampled systems suffer from aliasing errors (Young, 1974; Manfroid, 1985). The Johnson systems, both visible and IR, are seriously undersampled, and need

Table III. Extrapolation errors found by fitting Eq. (3) to Manduca and Bell's values from M = 0.0 to 3.0, assuming values of M_0 given in parentheses below each error.

	Kitt Peak		Johnson filters		
Δ (Vega)	J	K	J	K	L
Kitt Peak, summer	-0.005 (0.6)	-0.005 (1.6)	-0.006 (0.6)	-0.002 (1.0)	+0.004 (1.6)
Kitt Peak, winter	-0.022 (0.6)	-0.008 (1.6)	-0.030 (0.6)	-0.032 (1.0)	-0.012 (1.6)
Mauna Kea	-0.000 (1.0)	-0.005 (1.6)	-0.001 (1.0)	-0.010 (1.4)	
Δ (cool giant)					
Kitt Peak, summer	+0.010 (0.6)	-0.007 (1.6)	-0.012 (0.6)	+0.006 (1.0)	
Kitt Peak, winter	-0.030 (0.6)	-0.005 (1.6)	-0.033 (0.6)	-0.029 (1.0)	
Mauna Kea	-0.002 (1.0)	-0.001 (1.6)	+0.001 (1.0)	-0.011 (1.4)	
δ (giant - Vega)					
Kitt Peak, summer	+0.015	-0.002	-0.006	+0.008	
Kitt Peak, winter	-0.008	+0.003	-0.003	+0.003	
Mauna Kea	-0.002	+0.004	+0.002	-0.001	

additional bands inserted between the existing ones. Such bands, in the IR, would have the further advantage of providing more precise information on the water-vapor absorption, which should help in choosing good values of M_0.

It has long been known that the existing undersampled systems do not provide enough information to allow accurate correction for extinction, even in the visible. For example, Cousins and Jones (1976) state that "no equation involving B-V and U-B only will predict the extinction correctly for all luminosity types and different degrees of reddening.... Without more information, ... no rigorous colour correction is possible either for extinction or for colour transformation," and Blanco (1957), Mandwewala (1976), and Young (1974; 1988) obtained similar results.

All existing photometric systems violate the sampling theorem (Young, 1974). Thus, if we are ever to reach the full precision of the best observational data, we must develop adequately sampled photometric systems. Until this fact is generally recognized, however, it will continue to be impossible to obtain support for such efforts.

Other Considerations

The need to determine more extinction parameters in the IR than in the visible is a burden, but apparently an unavoidable one. The IR observer is forced to devote more observations to determining the extinction, because more parameters are needed to describe it, and these are not available as by-products of "color terms" as they are in the visible.

However, as shown in Young (1974), the extinction can be well determined with an acceptably small number of stars per hour if the observations are carefully planned. As the 3-parameter fits described above require observations at 3 different air masses, it is only necessary to increase the time devoted to extinction by 50%, compared to work in the visible. The largest M used should be about 2.5 or 3, beyond which the values become very uncertain owing to the variable scale height of water vapor. As in the visible, the determination of nightly extinction can be strengthened enormously by reducing several nights together (Young and Irvine, 1967; Manfroid and Heck, 1983).

To prevent aliasing time-dependent extinction into the fitted parameters, the observer should take care to observe both rising and setting stars and solve for time-dependent parameters, as in Rufener's (1964) "M and D" method. This is probably more urgent in the infrared than in the visible, because changes in water vapor can cause large effects on the IR extinction without producing any obvious visible effect. Furthermore, because the water vapor that dominates the IR extinction has a smaller scale height than the atmosphere as a whole, one should use an air-mass formula that takes this into account (cf. Schnaidt, 1938), as well as allowing for the distinction between true and refracted zenith distances.

Finally, it would be nice to derive the rational approximation for the Forbes effect theoretically from reasonable assumptions about the probability density distribution of monochromatic extinction within a band, rather than from the simple heuristic argument offered here. Band models may be helpful in selecting plausible distributions. It will also be useful for wideband work in the visible to cast the classical series expansions into the form of Eq. (3), using standard Padé methods, though this will be of little use to infrared workers.

References

Blanco, V. M. Band-width effects in photoelectric photometry. *Ap. J.* **125**, 209-212 (1957)

Cousins, A. W. J., and Jones, D. H. P. Numerical simulation of natural photometric systems. *Mem. R. Astron. Soc.* **81**, 1-23 (1976)

Forbes, J. D. On the transparency of the atmosphere and the law of extinction of the solar rays in passing through it. *Phil. Trans.* **132**, 225-273 (1842)

Guthnick, P., and Prager, R. Photoelektrische Untersuchungen an spektroskopischen Doppelsternen und an Planeten. *Veröff. Kgl. Sternwarte Berlin-Babelsberg* **1**, 1-68 (1914)

Hardie, R.H. in IAU Symp. 24, *Spectral Classification and Multicolour Photometry* (Academic, New York, 1966) p.243

King, I. Effective extinction values in wide-band photometry. *Astron. J.* **57**, 253-258 (1952)

Manduca, A., and Bell, R.A. Atmospheric extinction in the near infrared. *Pub.A.S.P.* **91**, 848 (1979)

Mandwewala, N. J. Analysis of Rufener's method for the atmospheric extinction reduction. *Publ. Obs. Genève* Ser. A, Fasc. 82 (Arch. Sci., Genève 29) 119-148 (1976)

Manfroid, J., and Heck, A. A generalized algorithm for efficient photometric reductions. *Astron. Astrophys.* **120**, 302-306 (1983)

Manfroid, J. "On photometric standards and color transformation" in I.A.U. Symposium 111, *Calibration of Fundamental Stellar Quantities*, edited by A. G. Davis Philip. (Reidel, Dordrecht, 1985) pp. 495-497

Rufener, F. Technique et réduction des mesures dans un nouveau système de photometrie stellaire. *Pub. Obs. Genève*, Sér. A, Fasc. 66 (1964)

Schnaidt, F. Berechnung der relativen Schichtdicken des Wasserdampfes in der Atmosphäre. *Meteorol. Z.* **55**, 296-299 (1938)

Strömgren, B. in *Handbuch der Experimentalphysik*, Band **26**, Astrophysik, edited by B. Strömgren (Akademische Verlagsgesellschaft, Leipzig, 1937) pp. 321-564

Young, A. T., and Irvine, W. M. Multicolor photoelectric photometry of the brighter planets. I. Program and procedure. *Astron. J.* **72**, 945-950 (1967)

Young, A. T. in *Methods of Experimental Physics,* Vol. **12** (Astrophysics, Part A: Optical and Infrared), edited by N. Carleton (Academic Press, New York, 1974) pp. 123-192

Young, A. T. in *Proceedings of the Second Workshop on Improvements to Astronomical Photometry* (NASA, 1988)

MODELS OF INFRARED ATMOSPHERIC EXTINCTION

Kevin Volk
Mail Stop N245-6
NASA Ames Research Center
Moffett Field, California, U.S.A., 94035

T. A. Clark and E. F. Milone
Department of Physics
University of Calgary
2500 University Drive N.W.
Calgary, Alberta, Canada, T2N 1N4

Abstract

Computer simulations of the atmospheric extinction are presented for the JHK filters. Comparison with observations show that the deviations from linearity in a magnitude/air mass plot are small even though the linear extrapolation to zero air mass produces a value much different than the models predict. The cause of this effect is discussed. It appears that direct verification of the non-linearity predicted by the models will be difficult to obtain.

Introduction

The question of how one should correct infrared photometric observations for the effects of atmospheric extinction, particularly for the medium-band systems that are generally used, is a problem that has been around since the extension of photometry from the optical wavelength range to infrared wavelengths. In the middle 1960's when the Johnson JKL... filter system was first established there was some question of whether the linear air mass/atmospheric extinction relation used for the UBV filter systems could also be applied at infrared wavelengths. In a paper which is rarely cited as far as the authors are aware, just before a well known paper in which the absolute calibration of the new filter system was presented, H. L. Johnson discusses whether one should use the linear extinction extrapolation to zero air mass to obtain the proper stellar magnitudes in the new filter system (Johnson, 1965). His conclusion was that because of strong molecular band absorption in the infrared the usual extinction correction methods would underestimate the extinction correction producing systematically high stellar magnitudes. Indeed, this conclusion can be traced back to at least the turn of the century especially in papers dealing with the solar infrared emission (see the references in Strong (1941) for example). To allow for this Johnson recommends extrapolation to an air mass value of less than zero in the correction for atmospheric extinction.

It is not clear whether this is what Johnson did when he was establishing his standard JKL... system. To be sure one would have to examine the original data for the various lists of standards published by Johnson and his associates. What is clear is that this idea was probably never used by others doing infrared photometry at that time and the matter was forgotten. 14 years later the whole question was re-discovered by Manduca and Bell (1979). Over the past 9 years the problem seems to have become generally known but no method for dealing with it has been found. As the accuracy of infrared photometry becomes better the problem becomes more pressing. Systematic effects at the level of a few hundredths of a magnitude to a tenth of a magnitude are now larger than the obtainable internal accuracy of the photometry; this is the magnitude of the non-linearity in the extinction.

To better understand the basic cause of the problem with the normal linear extinction relation it is useful to look at the basic physics behind the equation

$$m_\lambda(X) = m_\lambda(0) + k_\lambda \cdot X \qquad (1)$$

for the monochromatic magnitude $m_\lambda(X)$ observed for a source at an airmass X. $m_\lambda(0)$ is the magnitude that would be observed for the source above the atmosphere and k_λ is the absorption coefficient, at wavelength λ. Equation (1) is rather idealized as one never observes a true monochromatic magnitude. If written in terms of flux density equation (1) is just the standard law of absorption given in textbooks. Assuming that the ray path from the source passes through a homogeneous atmosphere this equation must apply at each wavelength.

One problem is that the atmosphere is not homogeneous. Strictly speaking different layers in the atmosphere will contribute differently to the extinction as a function of the air mass. Water vapour, found mostly low in the atmosphere, is subject to this effect. A more serious problem is that what are observed are integrated fluxes through a filter/detector system. If one takes equation (1) at each wavelength, writes out the expression for the observed source flux obtained by integration over a band-pass and then tries to convert back to a magnitude/air mass relation one quickly sees that in general it is not possible to recover an equation of the same form as equation (1).

That type of approach shows that for a given source the linear magnitude/air mass relation will hold for a broad-band filter if weak sources of absorption such as Rayleigh scattering and Mie scattering by aerosols are causing the extinction. It will also hold when weak line absorption is present. Different sources will have slightly different absorption coefficients as they average the k_λ values according to the shape of the spectrum. This introduces a colour dependence into the extinction equation.

When strong line absorption is present over the pass-band of the filter the linear magnitude/airmass relation must break down. Saturated spectral lines are no longer on the linear part of the curve of growth so doubling the column density will not double the absorption in magnitudes. The change in extinction is therefore less when X changes from 1 to 2 than it was in the original unit of column density and the extinction correction is underestimated. This is not a problem for the UBV filters because the extinction is almost entirely due to Rayleigh and aerosol scattering over that range of wavelengths.

The above arguments give some indication of the processes involved but they do not allow any quantitative conclusions to be drawn. The observational data at JHKL filter wavelengths should show this non-linearity in some way. Yet it must not be too extreme or it would have been noted right from the start. To find out what effects are expected one must construct some sort of model of the extinction for comparison with the data.

The Models

To study the effects of extinction for a number of sites under various conditions a series of atmospheric transmittance calculations were carried out. These calculations are similar to those of Manduca and Bell (1979). Most of the calculations were carried out for Kitt Peak with some being done for Mauna Kea, Calgary, and Mount Lemmon. Only the results for Kitt Peak will be discussed here. One important difference in these calculations is that newer JHK filter profiles than those considered by Manduca and Bell (1979) were used in the calculations. Another extension of the previous work was to test whether a model in which the atmosphere was considered as a set of layers rather than as a single layer makes any difference in the results. This section briefly describes the details of how the models were constructed.

There are a number of problems in calculating the transmittance of the atmosphere in the infrared. The absorption is mainly due to a large number of line transitions which have a wide range of strengths. The main molecules that provide the absorption are H_2O and CO_2. Various trace molecules are also significant sources of absorption. Water vapour and ozone are variable from site to site and over time; the other molecules can usually be considered as being constant constituents of the atmosphere. Matters are further complicated in that the line profiles are functions of pressure and temperature and so depend upon the structure of the atmosphere. Ideally each line would have to have a line shape calculated at a larger number of points along the ray path to obtain the correct absorption.

To allow for the inhomogeneity of the atmosphere the Curtis-Godson approximation was used to calculate mean absorption coefficients for the assumed atmospheric structure. This approximation defines an equivalent homogeneous layer which would give the same total line absorption in the weak-line and the strong-line limits as the assumed inhomogeneous path (see Goody (1964)). The Lorentz line shape was used in these calculations. The form of the Curtis-Godson approximation applied here assumes that the line strengths and the line profile are not strongly dependent upon the temperature so no corrections to the tabulated line parameters were done.

The atmospheric structure was assumed to follow that of the U.S. Standard Atmosphere (1976) mid-latitude summer model. The model was simply truncated at the appropriate altitude for the site under consideration. When the water vapour content of the atmosphere was changed in the various calculations the standard column density was scaled by a constant factor. Having set the physical conditions (column densities, temperature, pressure) for the model layer the calculation of the atmospheric absorption was carried out using the spectral line data tables of McClatchey et al. (1973). The compilation lists more than 100,000 lines of the molecules H_2O, CO, CO_2, N_2O, O_3, CH_4, and O_2 from a wavelength of 1 μm to the far infrared.

The atmospheric transmittance calculation was carried out from 3840 cm^{-1} to 10912 cm^{-1} (0.912 μm to 2.60 μm) with a grid spacing of 0.2 cm^{-1}. Actually the calculation was carried out at a slightly higher resolution and degraded to 0.2 cm^{-1} resolution since calculations at 0.002 cm^{-1} resolution for a series of test intervals indicated that the direct calculation at 0.2 cm^{-1} resolution was not sufficiently accurate. The 0.2 cm^{-1} grid spacing is of the same order as most of the line widths so the calculations were subject to a few percent rms error compared with the high resolution calculations. By doing a higher resolution calculation and degrading to 0.2 cm^{-1} the rms difference was reduced by a factor of between 5 and 10 depending upon which region of the spectrum was considered. The expected uncertainty estimated this way was about 0.005 magnitudes for a calculated flux value at 1.0 air masses.

The wavelength range used covers the J, H, and K filters obtained by Kitt Peak around 1979 for which filter profiles were provided by R. Joyce. These are not the same filters as were considered by Manduca and Bell (1979). For the stellar sources blackbodies of 9100 K and 4000 K were used. The former value matches the effective temperature of Vega near this wavelength region (Gehrz, Hackwell, and Jones 1974). For stars cooler than about 4000 K the molecular bands in the spectrum become important and a blackbody is no longer a good approximation to the spectrum. Finally, aerosol and Rayleigh scattering were calculated from the formulae of Hayes and Latham (1975) during the numberical integration to obtain the model magnitudes. To generate model magnitudes at different air masses the point by point atmospheric transmittance values were scaled according to equation (1). Air mass values of 0.1, 0.2, 0.3, 0.4, 0.5, 0.75, 1.0, 1.50, 2.0, 2.5, and 3.0 were used to produce the model extinction curves.

Results of the Models

In Figures 1 and 2 are shown two of the resulting sets of extinction curves for Kitt Peak using the single layer models. The first Figure shows a calculation for relatively dry summer conditions with 2.50 precipitable millimeters (pmm) of water vapour in the atmosphere. The curves are

Column Density Of H_2O (pmm)	Blackbody Source	Filter	E	Δ	Δ/E	δ
2.50	9100 K	J	0.069	0.067	0.970	+0.0026
		H	0.044	0.031	0.705	+0.0005
		K	0.051	0.031	0.597	−0.0009
	4000 K	J	0.070	0.070	1.007	
		H	0.044	0.031	0.717	
		K	0.051	0.030	0.587	
5.05	9100 K	J	0.085	0.099	1.174	+0.0036
		H	0.052	0.050	0.950	+0.0007
		K	0.059	0.037	0.624	−0.0007
	4000 K	J	0.084	0.103	1.221	
		H	0.052	0.050	0.965	
		K	0.058	0.036	0.624	
7.54	9100 K	J	0.093	0.119	1.284	+0.0037
		H	0.055	0.059	1.074	+0.0006
		K	0.062	0.039	0.640	−0.0007
	4000 K	J	0.092	0.123	1.333	
		H	0.054	0.059	1.087	
		K	0.061	0.039	0.636	
10.10	9100 K	J	0.100	0.137	1.363	+0.0037
		H	0.058	0.068	1.189	+0.0007
		K	0.065	0.043	0.660	−0.0006
	4000 K	J	0.099	0.140	1.412	
		H	0.057	0.069	1.202	
		K	0.065	0.043	0.657	

Table 1—Results of the single layer model calculations for Kitt Peak. The first column gives the water vapour column density for the models, the second column tells which blackbody was assumed for the source, and the third column gives the filter name. The last four columns give the numerical results: the slope of the extinction curve between 1.0 and 2.0 air masses, the zero point error from the linear extrapolation to zero air mass, the ratio of these quantities, and the difference in the zero point errors for the two blackbody sources. These quantities are denoted as E, Δ, Δ/E, and δ respectively. E and δ are in magnitudes and Δ is in magnitudes per air mass.

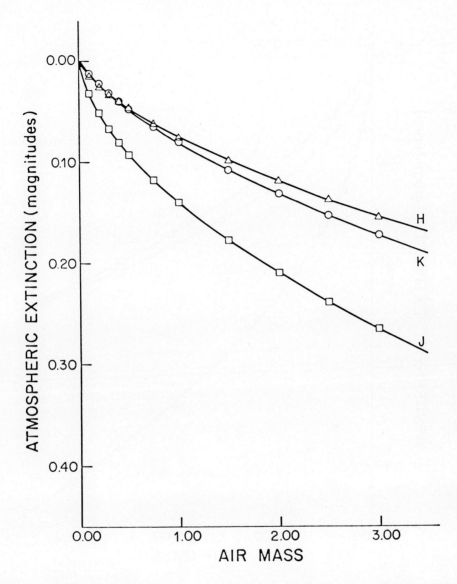

Figure 1: Example of the model extinction curves. This is for Kitt Peak with the 4000 K blackbody source and 2.50 pmm of water vapour.

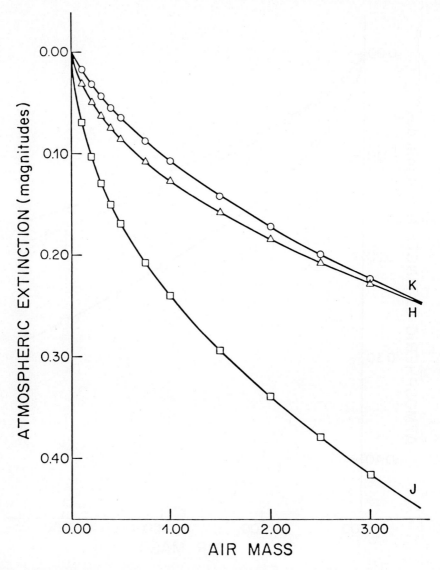

Figure 2: Another example of the model extinction curves, this time for 10.1 pmm of water vapour. The other parameters are the same as in Figure 1.

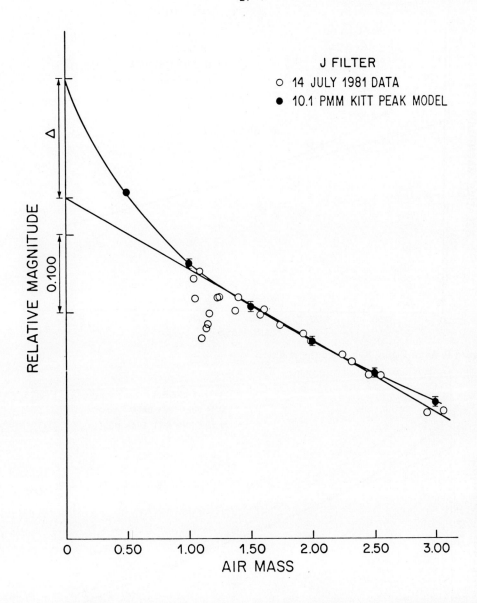

Figure 3: Comparison of the model extinction curve for the J filter with observations of Vega taken at Kitt Peak. The model is for 10.1 pmm of water vapour with the 9100 K blackbody. Data points are the open symbols. Model points are the solid symbols, joined by the curve. To produce this figure the model extinction at 2.0 air masses was matched to the data. Only relative values are illustrated in the plot so no y-axis scale marks are plotted. A straight line fit to the observed points is also plotted for comparison with the model curve.

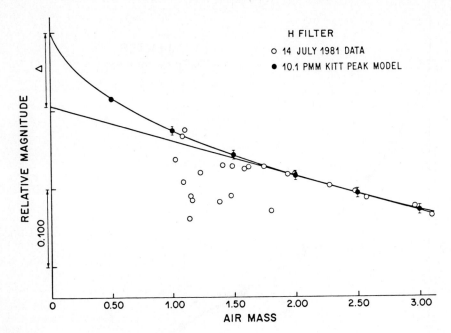

Figure 4: As in Figure 3, but for the H filter.

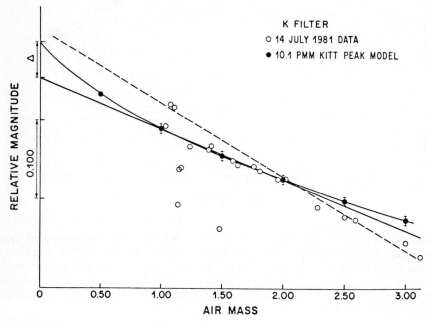

Figure 5: As in Figure 3, but for the K filter.

qualitatively similar to those of Manduca and Bell (1979). One sees that most of the curvature occurs at low X values where no observations can be made. The zero-point errors made from the linear extrapolation to zero air mass and the predicted slopes of the curves are given in table 1 for the 3 filters. The extent of the non-linearity is larger than that estimated by Johnson (1965).

In the second Figure is shown the results for 4 times as much water vapour on the same scales. The predicted slopes have increased, especially for the J filter, and the zero-point errors have also increased. Comparison of the figures shows a feature of the models which does not occur in Manduca and Bell (1979): the H and K extinction curves are predicted to cross over at some X value which increases as the water vapour column density increases. This seems to be due to different molecular bands affecting the filters as the air mass increases. Due to the newer filters these model curves are somewhat different than those in Manduca and Bell (1979).

Both the figures are for the 4000 K blackbody source. In all of the models the colour effects were found to be relatively small. The results are given in Table 1 using the notation of Manduca and Bell (1979) for the slope, the zero-point error, and the colour effect between the 9100 K and 4000 K blackbodies.

Some models were also run in which the atmosphere was divided into 8 layers to more accurately allow for the distribution of water vapour in the atmosphere. The overall column density values were kept the same as for the single-layer models for comparison. It was found that the single layer models give too large an extinction at X values greater than 1, but that this effect is rather small. The values at 1.0 air masses differ by less than 0.002 magnitudes between the two calculations. It appears that this difference is due to the strong lines which saturate high in the atmosphere so that a single layer model overestimates the line wing absorption slightly. This would mean that the line parameters have a large enough temperature dependence to be noticeable and the Curtis-Godson approximation as used here is not quite correct. The effect is small enough to be neglected considering the other uncertainties in the models. Something which should be done in conjunction with such a multiple-layer model is to explicitly calculate the raypath through the various layers, particularly if the calculations are to be extended to large air masses. This could be a more serious effect than the difference between the multiple-layer and single-layer models found in this study.

It is important to compare the models with actual data to see if the effects of the non-linearity should be observable. For that purpose comparison was made with observations of Vega taken on the night of 14 July 1981 at Kitt Peak. To do this the slope of the observed magnitude/air mass relation was compared to the models. It was found that the 10.1 pmm model had a slope almost identical to that in the data. On the night the observations were taken there were thunderstorms in the vicinity of Kitt Peak and thus it is not surprising that conditions were rather wet. To carry out detailed comparisons for the 3 filters the model extinction curves were normalized to the data at 2.0 air masses. The results of this are shown in Figures 3 to 5.

Looking at the data there is considerable scatter in the values for air masses between 1.0 and 1.5 but at larger air masses the extinction had settled down, presumably as the thunderstorms had gone away, and the extinction relation seems nicely linear. In Figures 3 and 4 a line has been drawn based upon the points beyond 1.5 air masses. In Figure 5, the K filter results, it was not clear how to draw in a line because the extinction relation seems to have a change in slope around 2.0 air masses. As a result, two lines were drawn in that case.

In all three cases the slope from around 2.0 air masses in the data is very similar to that predicted by the models. For the K filter results there is a distinct difference between the data values for $X > 2.0$ and the model prediction. In the other 2 cases the match is very good out to 3.0 air masses. For the J and H filters there are points near 1.0 air masses which agree with the model and so appear to be values where the observations are not effected by the clouds which caused problems for most of the 1.0 to 1.5 air mass range. That is not the case for the K filter where the corresponding points are above the model values. Possibly the lack of agreement for the K filter indicates that the model calculations for this filter are incorrect for some reason.

The figures show how subtly the model curves deviate from a straight line in the region between 1.0 and 3.0 air masses even though there is a significant difference in the zero point values. This is for rather wet conditions when the non-linearity is strong. Under good conditions it seems that there would not be observable curvature at the 0.01 magnitude level of photometric accuracy for the J filter. Even at 3.0 air masses the deviation between the model curve and the data values is barely 0.01 magnitudes.

In the case of the H filter it is possible that a small amount of curvature has been observed near 1.0 air masses. Again this curvature is barely as large as the accuracy of the observations and would not be detected by ordinary data reduction techniques. One has the unfortunate result that the curvature can only be observed under very wet but very stable atmospheric conditions. That combination is unlikely to be found very often.

Another conclusion from this comparison is that even if one can produce models specific for the site of observation it will be difficult to directly verify the model of the extinction. It may be possible to empirically study the slope of the observed extinction curves and the zero-point magnitudes obtained. From the models these quantities should be correlated in a predicable way. The usual data reduction techniques absorb the change in the zero-points into the transformation the standard system. Studies such as that of Glass and Carter reported at this meeting show the type of effects that the models predict and may therefore allow the verification of the models. That would allow a set of standard extinction corrections to be derived from the models which could be easily implimented. Due to the very small difference between the model curves and a straight line over the $X = 1.0$ to $X = 3.0$ range the direct fitting of model curves to the data would be rather unstable.

References

Gehrz, R. D., Hackwell, J. A., and Jones, T. A. 1974, *Ap. J.*, **191**, 675.
Goody, R. M. 1964, *Atmospheric Radiation*, (Oxford: Clarendon).
Hayes, D. S. and Latham, D. W. 1975, *Ap. J.*, **197**, 593.
Johnson, H. L. 1965, *Comm. Lun. Plan. Lab.*, **3**, 67.
Manduca, A. and Bell, R. A. 1979, *Pub. Astr. Soc. Pac.*, **91**, 848.
McClatchey, R. A., Burch, D. E., Rothman, L. S., Benedict, W. F., Calfec, R. F., Garing, J. S., Clough, S. A. and Fox, K. 1973, *AFCRL Atmospheric Absorption Line Parameters Compilation*, AFGRl-TR-73-0096, Bedford, Mass.
Strong, J. 1941, *J. Franklin Inst.*, **232**, 1.
U.S. Standard Atmosphere 1976, NASA, USAF, USWB, (Washington D.C.: U.S. Government Printing Office).

Discussion:

Q. John Graham (Carnegie DTN) - How dependent are your calculations on the precise JHK filter pass-bands used?

A. The model results are sensitive to the filter profiles because the filters are in narrow atmospheric windows and so slight shifts in the filter can bring in strong band absorption on the edges of the window thereby changing the extinction. The detailed values will change.

ATMOSPHERIC EXTINCTION IN THE INFRARED

Ronald J. Angione
Astronomy department, San Diego State University
San Diego, California 92182-0334

Abstract

Atmospheric extinction due to Rayleigh, water vapor, and mixed gases is discussed. Particular emphasis is given to water vapor absorption. A method for determining the amount of precipitable water vapor, calibrating it using LOWTRAN, and calculating the atmospheric transmission has been in use at Mount Laguna Observatory for five years. This method and the results from it are discussed.

Introduction

We have a long-term program monitoring the solar constant at Mount Laguna Observatory (located 40 miles east of San Diego at an elevation of 1860 meters). The total solar irradiance reaching the ground is measured by an absolute cavity radiometer, of which we have three for intercomparison. In order to determine the solar constant we must measure the atmospheric transmission, model the light loss from the UV to the far IR, and add this lost energy to the ground-measured value to obtain the solar constant. It is the atmospheric transmission determination that is of interest here.

Figure 1 shows an overview of the basic atmospheric measurement and modelling procedure. Atmospheric extinction can be divided into three components: molecular scattering, aerosol scattering, and molecular absorption. These last two components show considerable variability and must be measured at the time of program object observation. In our program we use a filter-wheel radiometer to measure the amount of aerosol, water vapor, and ozone; the remaining mixed gases are scaled be barometric pressure. We then model the atmospheric transmission from 0.25 to 28.5 microns.

The filter-wheel radiometer is similar to a standard astronomical photometer. A Fabry lens images the front, 5 mm aperture onto the EG&G UV 444B silicon photodiode detector. A twelve position filter wheel lies in front of the detector. The filters are all interference filters with a FWHM of typically 75 Angstroms. Eight filters are placed in atmospheric absorption clear spectral regions, two in the Chappuis ozone band, and one in the 0.935 micron water vapor band. The radiometer has an

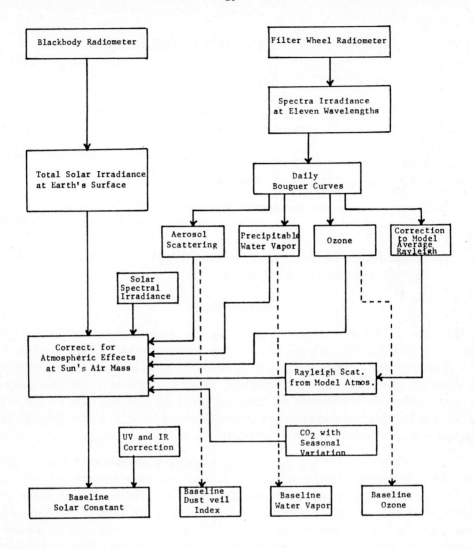

Figure 1. The atmospheric and solar measurement scheme at Mount Laguna Observatory.

unvignetted field of view of one degree and uses an active tracking system to automatically follow the sun to better than one arcminute. Both the filters and detector are temperature controlled to better than 0.5 degrees Celsius year round, and the instrument is kept under positive pressure of dry nitrogen. The radiometer is completely automated, except that an observatory staff member must roll off the protective building, point the instrument in the vicinity of the sun, and close up in the afternoon.

Although this instrument looks at the sun, the same techniques should be applicable to night-time observations. The atmospheric transmission results for the morning hours should also be similar to night-time measurements.

Rayleigh Scattering

Molecular extinction, or more properly Rayleigh-Cabannes scattering, is both a reasonably well understood extinction component, and a minor contributor to the extinction in the infrared.

Given for each gas component of the atmosphere (1) its fractional volume, (2) its index of refraction, and (3) its depolarization factor you can calculate the scattering coefficient for a particular monochromatic wavelength. A frequently cited reference is Penndorf (1957). Several formulas have been developed for Rayleigh scattering by the Earth's atmosphere (e.g. Hayes and Latham, 1975), which can be scaled by the local atmospheric pressure at the time of observation. I use a formula by Frohlich and Shaw (1978), which must be corrected for the proper depolarization factor (see A. T. Young, 1980). The Rayleigh extinction at our observatory altitude is only 0.003 magnitudes/airmass at 1.25 microns (J-band). Atmospheric pressure changes would cause these values to change by only about one percent.

Aerosols

Extinction due to aerosol (dust) particles, called Mie scattering, is significant and variable in the IR region covered by the JHK filters. It takes approximately 2000 calculations to determine the intensity of light at just one monochromatic frequency scattered at one angle by a spherical particle of one radius and index of refraction. Empirical formulas are in widespread use! The most often used formula for extinction due to aerosol scattering is:

$$k_a = B \lambda^{-N} \qquad (1)$$

At eight wavelengths from 0.38 to 1.01 microns, chosen to avoid telluric absorption features, the total extinction is determined from the slopes of Bouguer curves (Δmag. versus airmass). Figure 2 shows a Bouguer plot for a wavelength of 1.01 microns. Note that each horizontal scale mark is 0.001 magnitudes, indicating that the precision of a single measurement is considerably better than this. The Rayleigh component is calculated and subtracted to leave just the aerosol. A linear least squares fit to the logarithm of equation 1 gives both B and N. Knowing B and N allows one to calculate the aerosol extinction at any wavelength. If you know N, then you could determine B by making a measurement at just one wavelength, 1 micron for example.

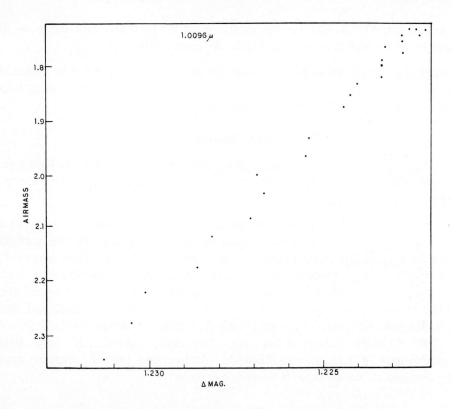

Figure 2. Bouguer plot obtained with filter wheel radiometer showing the precision that can be obtained for measuring aerosols.

However, the optical properties of the aerosol particles depend on their size distribution. This is further complicated because different kinds of particles (e.g. sea salts, hydrocarbons, volcanic debris) should have different size distributions. Empirical studies of the number, n, of particles with radius, r, larger than 2 microns gives a distribution

$$\frac{dn}{dr} = r^{-m} \qquad (2)$$

with m approximately 2 (Toon and Pollack, 1976). This gives the familiar $1/\lambda$ wavelength dependence for aerosol extinction. When one includes the smaller particles, and their absorption component, then the aerosol extinction becomes more complicated. The work by Toon and Pollack (1976) shows that the aerosol extinction decreases smoothly with increasing wavelength to about 2.5 microns. Beyond 2.5 microns it becomes complex with numerous absorption peaks due to the particles such that the aerosol component of extinction at 10 microns can nearly equal that at 1 micron.

I use, instead of equation 1, the following empirically determined relation for the aerosol extinction.

$$k_a = B\lambda^{-(N + C\log\lambda)} \qquad (3)$$

where B, N, and C are coefficients determined by least squares. The aerosol component is also highly variable, showing changes of 50% in just a few days, even during normal atmospheric periods. During volcanically perturbed periods the aerosol component can more than double along with dramatic changes in the wavelength exponent, as was observed at Mount Laguna after the eruption of El Chichon (4 April, 1982). Observed values of the aerosol extinction component at 1.01 microns, in units of mag./airmass, are presented in Table 1, along with the $1/\lambda$ extrapolations to the J, H, K, pass bands. The volcanic aerosols are stratospheric, so even being on Mauna Kea will not help. It should also be noted that the aerosol extinction changes by about 50% across these filters.

Table 1

	1.01 microns	J	H	K
Max.	0.057	0.046	0.035	0.026
Min.	.005	.004	.003	.002
Mean	.020	.016	.012	.009
After the eruption of El Chichon				
Max.	.152	.123	.093	.068
Min.	.036	.029	.022	.016

Precipitable Water Vapor

The technique used for monitoring water vapor is described in detail by Angione (1987), and the results are compared to those at Kitt Peak (Wallace and Livingston, 1984; Wallace, et al., 1984). This simple, straight forward method is essentially the same as that developed by Fowle (1912 and 1913) for the Smithsonian solar constant program; however, our method of calibration is quite different. In our measurements we determine the depth of the 0.94 micron water vapor band using the ratio of intensities through a filter in the band to that outside the band. The "continuum" filters are centered at 0.8507 and 1.0097 microns. All filters have a FWHM of nominally 75 angstroms. The ratios of intensities I(8507)/I(9457) and I(10097)/I(9457) each give a determination of the precipitable water vapor.

The difficult part is the calibration of this ratio in terms of precipitable water vapor, which we do spectroscopically. A coelostat sends sunlight to a Spex model 1672 grating double monochromator, with a temperature-controlled silicon photodiode detector, and the precipitable water vapor is determined simultaneously with the filter measurements. Figure 3 shows a tracing of the 0.94 micron water vapor band along with the position and profile of our water vapor filter superimposed. Although I use the sun, the same procedure would work at night (1) using a CCD equipped spectrograph on a star, or (2) calibrate the filter ratios using the sun, and then use the filters at night on a G-type star.

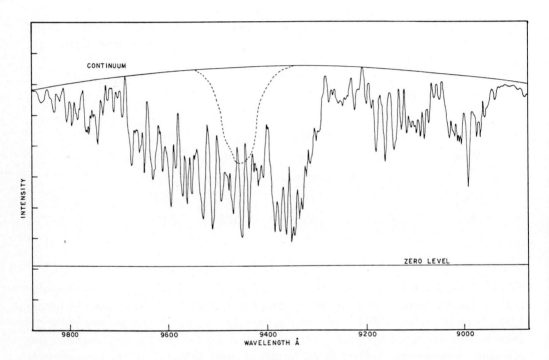

Figure 3. Tracing of the 0.94 micron water vapor band at Mount Laguna with the inverted filter response shown as a dashed line.

The band area is determined by planimetry. To calibrate the band area in terms of water vapor I used the LOWTRAN 2 program (Selby and McClatchey, 1972) to generate synthetic spectral plots for different known amounts of water vapor. Although more advanced versions of LOWTRAN (e.g. LOWTRAN 6) are available, the water vapor is determined in the same way. The synthetic band areas were also determined by planimetry. Figure 4 gives the relationship between band area and water vapor, with the estimated precision of a single point 3%.

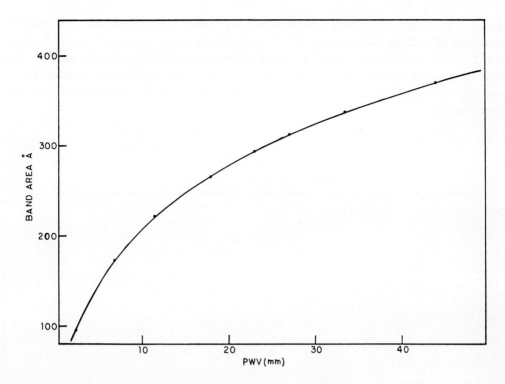

Figure 4. Relationship between band area and precipitable water vapor determined from synthetic spectra using LOWTRAN.

LOWTRAN is a low resolution (20 cm^{-1}) transmission code for 0.25 to 28.5 microns developed by the AFCRL. It divides the atmosphere into 33 layers from 0 to 100 km, specifying the temperature, pressure, and composition for each layer. For a given zenith angle the optical path is is refracted through each layer and the optical depths added up. This it does very well and takes into account the different scale heights for the mixed gases, water vapor, ozone, and aerosols. The absorption coefficients were determined separately for each gas from laboratory measurements of transmittance (McClatchey, 1972). For most bands exact analytical expressions are used, but the strong and weak line approximations of Goody (1964) are also used. However, because the absolute accuracy of the absorption coefficients is only about 5% (Howard, 1956), LOWTRAN will not calculate the transmission well enough to use for photometric reductions.

The final step in this calibration is to determine the relationship between the measured water vapor and filter ratios. Figure 5 shows this relationship for one of the two ratios. The relationship is well approximated by a linear least squares fit. The standard deviation of a single measurement is 0.7 mm, and the mean error is 0.2 mm. In practice the water vapor is calculated from both ratios, printed out separately as a check, and then the average value is used. A correction is made for the differential continuum optical depth between the intensity at the continuum points and the intensity at the water vapor band.

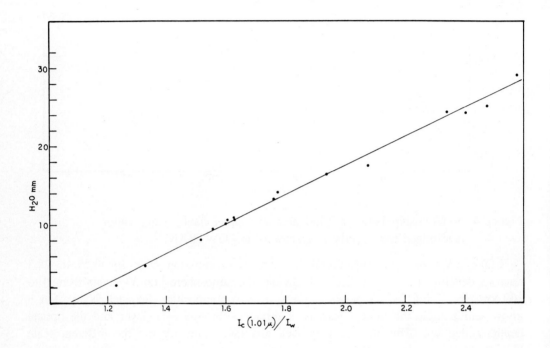

Figure 5. Empirical relationship between precipitable water vapor and filter ratio at 1.01 microns for Mount Laguna.

Figure 6 presents a time series of the measured water vapor, scaled to one zenith airmass, above Mount Laguna. The values of water vapor presented here were measured at about 0930 PST and should therefore be indicative of the water vapor during the previous night. The only obvious features are the pronounced seasonal variation and the large night to night variations. The seasonal variation is shown in the modulo one year Figure 7.

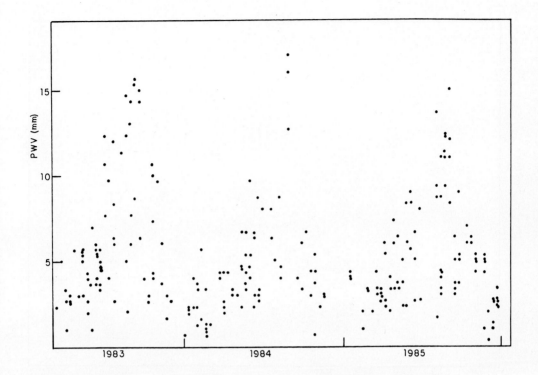

Figure 6. Time series plot of water vapor measured at Mount Laguna.

In an earlier study (Roosen and Angione, 1977) of a larger, but older data set from the Smithsonian solar constant project, we found results consistent with modern measurements. Drawing upon this larger data base, Figure 8 shows the long-term variability of water vapor for the two main Smithsonian sites Mount Montezuma, Chile and Table Mountain, California. The curves are twelve-month running means, so the seasonal variation has been smoothed out to show the long-term changes.

Conclusion

Proof that this method of water vapor, aerosol, and Rayleigh scattering determination and correction works is that, although I must add a total atmospheric correction of over 25%, my mean value for the solar constant is 1365 compared to 1368 watts/meter^2 measured by the SMM satellite experiment. The standard deviation for a single day is 0.3%, or approximately 0.003 magnitudes.

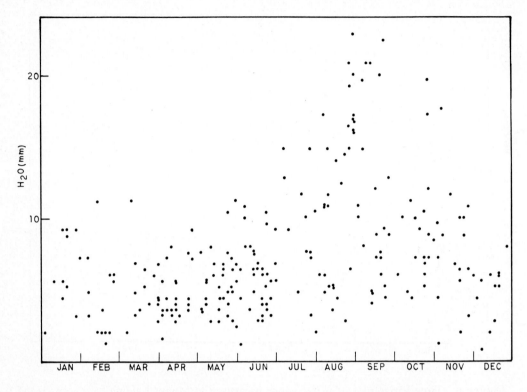

Figure 7. Modulo one-year water vapor at Mount Laguna, illustrating the seasonal variation.

Water vapor is such a highly variable quantity, and is so important to the extinction in the infrared, that if one needs to know it, then it must be measured nightly. This is also true for the aerosol component, particularly during volcanically perturbed times.

The method described in this paper works for my purpose because I am integrating over the whole spectrum convolved with the solar spectral energy distribution. I do not think it would work for standard astronomical photometry until better values of the absorption coefficients are known. Once those coefficients are known, it should be possible from (1) measuring the barometric pressure to scale the Rayleigh and mixed gas component, (2) measuring the extinction at 1 micron to get the aerosol component, and (3) measuring the water vapor, to calculate the extinction for the J, H, K bands more accurately than it can be determined from nightly airmass plots.

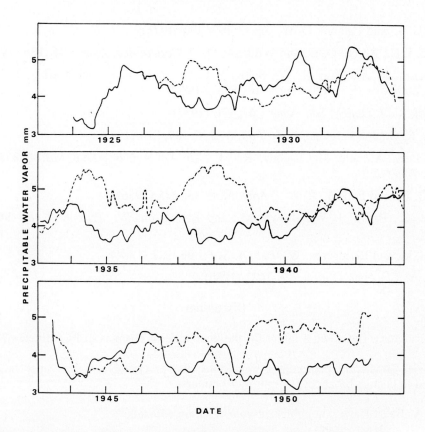

Figure 8. Long-term water vapor variations at Mount Montezuma, Chile (solid line) and Table Mountain, California (dashed line) from the work of Roosen and Angione (1977). These plots are twelve-month running means to show the long-term variations.

References

Angione, R. J. *Pub. A. S. P.* **99,** 895 (1987)

Fowle, F. E. *Ap.J.* **35,** 149 (1912)

Fowle. F. E. *Ap.J.* **37,** 359 (1913)

Frohlich, C., and Shaw, G. E. *Phys. Meteorol. Obs., World Rad. Center Pub.* **No.**557, (1978)

Goody, R. M. in *Atmospheric Radiation, Vol. 1, Theoretical Basis* (Oxford University Press, London, 1964)

Hayes, D. S., and Latham. D. W. *Ap.J.* **197,** 593 (1975)

Howard, J. N., Burch, D. E., and Williams, D. *J. Optical Soc. Amer.* **46,** 186 (1956)

McClatchey, R. A., Fenn, R. W., Selby, J. E. A., Volz, F. E., and Garing, J. S. *Air Force Cambridge Res. Lab. Report AFCRL-72-0497*

Penndorf, R. *J. Optical Soc. Amer.* **47,** 176 (1957)

Roosen, R. G., and Angione, R. J. *Pub. A. S. P.* **89,** 814 (1977)

Selby, J. E. A., and McClatchey, R. M. *Air Force Cambridge Res. Lab Report AFCRL-72-0745*

Toon, O. B., and Pollack, J. B. *J. Appl. Met.* **15,** 225 (1976)

Wallace, L., Brault, J. W., Brown, M., and Livingston, W. *Pub. A. S. P.* **96,** 836 (1984)

Wallace, L., and Livingston, W. *Pub. A. S. P.* **96,** 182 (1984)

Young, A. T. *Appl. Optics.* **19,** 3427 (1980)

Discussion

"Q (F. Rufener): Have you a figure for the time-scale of changes in IR extinction?"

"A: Under apparently photometric conditions, the precipitable water vapor has been observed to change by about 0.5 mm in 20 minutes."

"Q (R. A. Bell): Have you published your work on aerosols?"

"A: Some earlier work has been published on analysing the Smithsonian data (Roosen, Angione, and Klemcke, 1973, Bull. Amer. Met. Soc. **54,** 307). Recent work is in preparation for publication."

INFRARED EXTINCTION AT SUTHERLAND

I S Glass & B S Carter
South African Astronomical Observatory, P O Box 9,
Observatory 7935, South Africa

Summary

Measurements of the extinction coefficients per unit airmass in the JHKL (1.25, 1.65, 2.2 and 3.5µm) bands are presented. The nightly values of the photometric zero-points have also been analysed. It is shown that there is large scatter and an annual variation in the J-H and K-L zero-points while that for H-K is very stable.

Introduction

JHKL photometry has been conducted at Sutherland (alt. 1760m, lat 32° 23'S) since 1972. In general, it has not been our policy to determine the extinction coefficients nightly but rather to use the "default" values (E_J, E_H, E_K, E_L) = 0.10, 0.06, 0.10, 0.15) which were those in use at Mts Wilson and Palomar when we started and which appeared adequate for our site. Errors resulting from variations in the assumed extinction were neglected (a) because we usually worked between 1 and 1.3 airmasses and (b) because the errors in the assumed values of the standard stars probably ranged up to around .03 mag. However, Carter (1984) has presented a new standards list of much greater accuracy than our previous one, having errors of .02 mag or better (Glass, 1985) at J,H and K and somewhat greater than this at L. To take full advantage of the improved accuracy, a more careful treatment of the extinction is required. At the same time, the problems of determining the extinction coefficients have been eased.

Extinction Coefficients

The accuracy with which the extinction coefficients can be determined is limited: on nights when the atmospheric transmission is high the effects are small and hard to measure; on nights of high extinction errors often arise from the general instability of conditions.

Figure 1 shows the extinction curves for an exceptionally dry and stable night (23-24 February 1987) when a red and a blue star were followed up from the horizon. The J,H and K curves are very well behaved, having scatter of < .01 mag whereas the L curve has a scatter

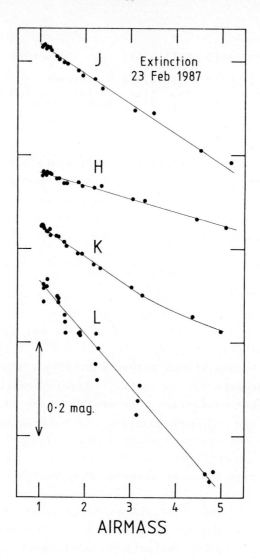

Figure 1. Extinction vs. airmass on 23 February 1987.

of around 0.03 mag. Only the blue star is illustrated in Fig 1: no difference is obvious for the red one. The J and H relations in fact appear to be straight lines whereas that for K curves slightly. The K extinction per unit airmass is lower for higher airmasses (3-5) than for lower ones (1-3), although the curvature is negligible for airmasses less than 3.

Figure 2 shows extinction curves from a typical night (27-28 February 1988) and illustrates one of the problems which has to be faced in this work. Measured at the zenith, this night appeared to be stable within ~ .01 mag during the observations. Between 1 and

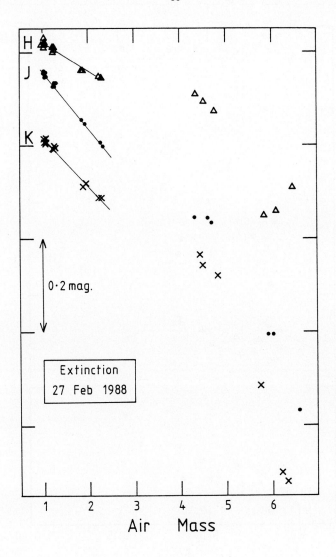

Figure 2.   Extinction vs. airmass on 27 February 1988.

2.3 airmasses the results are consistent but if slopes are determined from the photometry at higher airmasses the results vary by almost a factor of two. Although the measurements were derived from observations of several different stars on the night in question, this fact cannot account for the total scatter. It is believed that the fluctuations are due to atmospheric layering and they show the necessity of making determinations of airmass coefficients near the airmasses of the programme objects; i.e. between 1 and 2.

Figure 3 summarizes our information on extinction coefficients from several different nights, plotted in the form E_λ vs E_K.

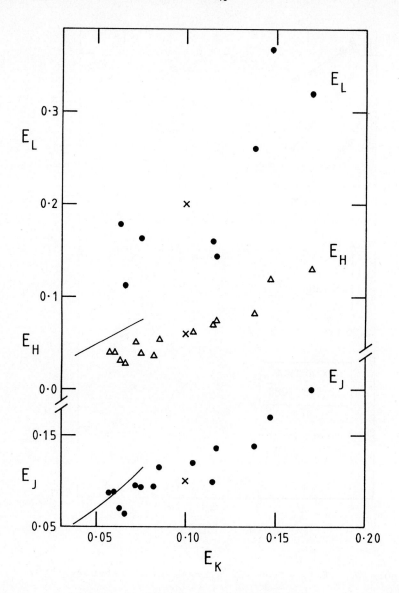

Figure 3. Extinction coefficients at J, H and L plotted against that at K for several nights. The night of 19 September 1986 has been divided into two parts.

Note that E_H is closely proportional to E_K : $E_H = 0.80\ E_K - .02$. Similarly, $E_J = 0.95\ E_K + .02$ but with more scatter. The E_L points are so scattered that no definite relationship with E_K can be specified. It is hoped that a clearer trend will be discernable when more data have been accumulated. The figure also shows two theoretical lines from the work of Manduca & Bell (1979) for filters having closely similar transmissions to ours (the "Kitt Peak" set).

The extremes of these lines are for 1.7 and 10.1 mms precipitable water vapour. While the predictions are in the right general range, the K extinction seems to have been under-estimated.

Zero Points

The zero-points of our photometry which, of course, mainly depend on extinction within airmasses 0 to 1, vary considerably from night to night. Fig 4 shows two examples of 1-week observing runs with the zero-points for each night plotted sequentially. It is conspicuous that all bands vary together but that J and L have much greater amplitudes, presumably reflecting the fact that the latter bands are much

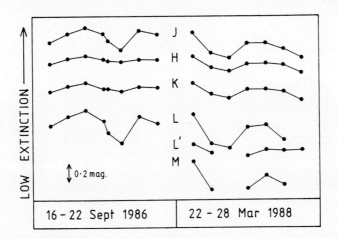

Figure 4. Variations of nightly average zero-points during two observing runs of one week each. The night of 19 September 1986 has been divided into two parts.

more affected by water vapour. The advantage of using L' (centred at 3.8μm) instead of L is shown in the March 1988 data. As expected from the distribution of water-vapour absorption bands, the variations at M are usually even greater than those at L.

The long-term changes in the magnitude zero-points unfortunately depend on how exactly our detectors are "peaked up" each time the instruments are placed on the telescopes or when a new observer takes over, which is usually weekly. However, the colour zero-points are expected to be independent of this. Fig 5 shows a year's worth of average nightly zero-points for J-H, H-K and K-L. About 250 nights were used to make up this diagram including a fair number when photometry was barely possible.

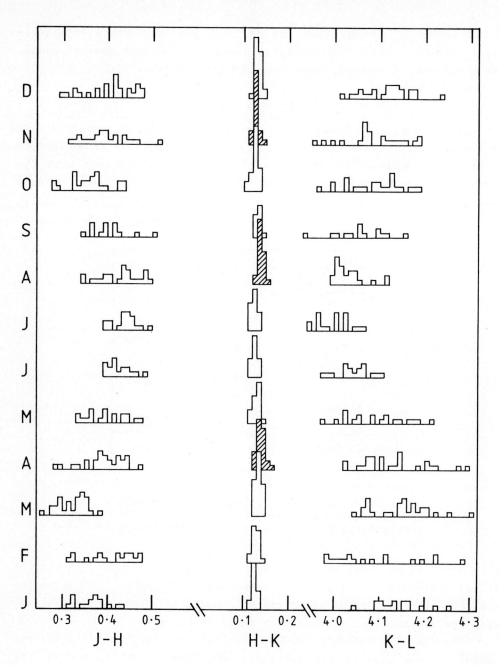

Figure 5. One year's worth of nightly colour zero-points from work at the 0.75m telescope in 1987. Photometry was attempted on about 250 nights, somewhat more than usual because of SN1987A. It is thus possible that nights of high atmospheric humidity are over-represented. The telescope was re-aluminized during February but the effects of this are believed to be negligible, based on the stability of the H-K zero-point.

The H and K zero-points vary closely together so that H-K has an annual range of only about .01. The J-H and K-L values show considerable scatter and an annual trend. J-H reaches a maximum and K-L a minimum during our winter months (June, July and August) when the water vapour content of the atmosphere is lowest due to lower temperatures. The scatter and annual range of the K-L colour are higher than the same for J-H.

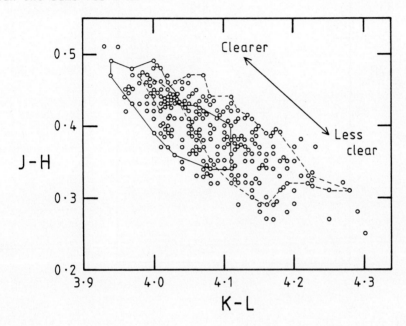

Figure 6.   J-H vs K-L zero-points for each night in 1987.   The summer points lie within the dashed line and the winter ones within the solid line.

In Fig 6 the J-H and K-L zero-points for each night are plotted. The outlines show the areas occupied by the summer points and the winter ones when taken separately. If one regards the longitudinal spread as due to water vapour content it is clear that a second parameter is also at work. This is very likely to be temperature. No obvious seasonality is apparent in the optical region zero-points measured at Sutherland (J W Menzies, private communication) and this appears to preclude dust content of the atmosphere as the responsible variable. From limited data, the relative humidity near ground level does not appear to be a good general predictor of the extinction although, of course, they rise together before the advent of low-level clouds.

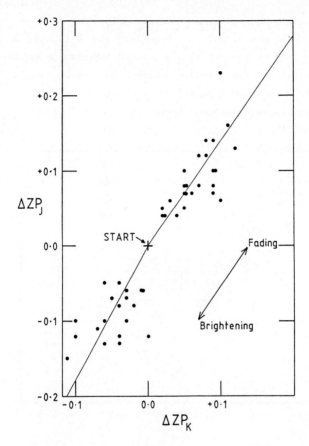

Figure 7. Nightly ΔZP_J vs ΔZP_K. The cross marks the start of the night.

The correlation of the zero-point changes from band to band is also of great interest. Because the zero-points are liable to change due to variations in the setting-up, the long-term changes cannot be determined. However, we have examined about 100 individual nights of 1.9m photometry for variations in the residuals of the standard stars. About 50 nights had shifts of > .03 mag in one or more of the JHK zero-points. These shifts are shown in Figs 7, 8 and 9. As expected, the ΔZP vs ΔZP graph is close to a straight line; higher scatter is associated with nights when the changes were greatest and the overall accuracy of the photometry was presumably lower. Note that ΔZP_J and ΔZP_L, plotted against ΔZP_K, show greater scatter than ΔZP_H.

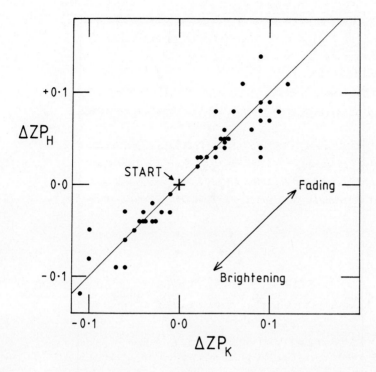

Figure 8. Nightly ΔZP_H vs ΔZP_K. The cross marks the start of the night.

The slopes of these graphs show that the ratios of the colour changes are not the same as the values found for the changes in extinction coefficients. $\Delta J/\Delta K$ is about 1.3 times higher here and $\Delta H/\Delta K$ is about 1.6 times more. Thus it is seen that the changes in the "first airmass" are not strictly proportional to those in the "second".

It would be very useful if the easily-determined colour zero-points could be used to predict the extinction coefficients. So far, the observational evidence suggests that there is at least a correlation but the amount of scatter in the relation has still to be determined. It will be necessary to obtain extinction coefficients for a large number of nights to examine this question properly. Fig 10 summarizes the data available at present.

Finally, let us look at the range of zero-point variation which can take place in only a few hours. Fig 11 gives the data for 20-9-86 which shows some of the most extreme variations we have encountered. At present, for nights such as this one, we plot the magnitude zero-points versus the time of observations of each standard

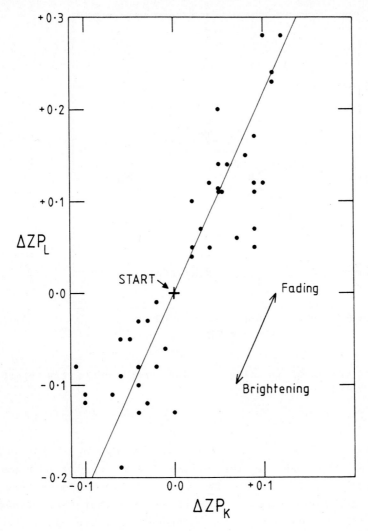

Figure 9.   Nightly ΔZP_L vs ΔZP_K.   The cross marks the start of the night.

star and interpolate by hand.   The extinction coefficients cannot be determined for every moment during the night if programme objects are to be observed, so we use the default values.   The accuracy of our photometry is preserved by making use of standard stars close to the objects being investigated and having as similar a colour as possible. Clearly if a relationship between the extinction coefficients and the colour zero-points can be established, the overall accuracy of the photometry can be improved.

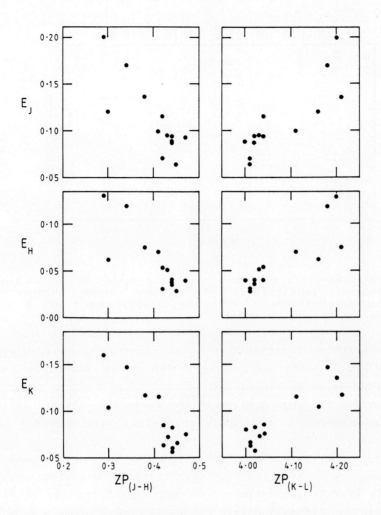

Figure 10. Plots of E_J, E_H and E_K against the (J-H) and (K-L) colour zero-points.

References

Carter, B.S., 1974.   M.Sc. Thesis, University of Cape Town.
Glass, I.S., 1985.   <u>Irish A.J.</u>, 17, 1.
Manduca, A. & Bell, R.A., 1979.   <u>Publ. Astr. Soc. Pac.</u> 91, 848.

Discussion

Q. (M. Bessell): Your new J filter transmits further to the blue than did your older filter.   Is the extinction and ZP of the new filter much worse than the old one?
A: We have not noticed any significant difference but have not looked

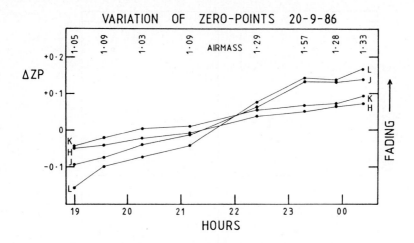

Figure 11. Variation of standard star residuals during a night showing particularly strong and rapid changes (20 September 1986). Note how closely the H and K zero-points track each other.

into this question systematically. The transformation $(J-H)_{Carter}$ = 1.05 $(J-H)_{Glass}$ works well except for miras and other stars heavily affected by molecular bands.

Q. (J. Koorneef): You show a slide of the zero-point changes during a (infrared-) poor night. How could that night have appeared to an optical observer?

A: The optical observer would have had to close down at the same time as the infrared one! From the fact that there is no obvious seasonal effect on the visible extinctions I do not expect them to be highly correlated with the infrared ones.

NEAR-INFRARED EXTINCTION MEASUREMENTS AT THE INDIAN OBSERVATORY SITES

N.M. Ashok
Physical Research Laboratory
Ahmedabad 380 009, India

ABSTRACT

Regular near-infrared astronomical measurements have been made since 1978 at the Uttar Pradesh State Observatory, Nainital (79°.46 E; 29°.36 N) and the Indian Institute of Astrophysics Observatory, Kavalur (78°.83 E; 12°.58 N) using liquid nitrogen cooled InSb photometers developed at the Physical Research Laboratory (PRL), Ahmedabad. Using the observations of the standard stars from the list of Johnson et al (1966) the JHKL photometric system of PRL has been tied to the broad-band infrared photometric system defined by Johnson (1965, 1966). The average J, H and K band extinction coefficients are 0.15 mag, 0.07 mag and 0.09 mag respectively. These are in general agreement with the theoretical extinction coefficients calculated by Manduca and Bell (1979) for 5 mm atmospheric water vapour and the Kitt Peak J H K filters. The details of the transformation from instrumental system to standard system are also dicussed.

I. Introduction

The ground based near infrared astronomy programme was initiated at Physical Research Laboratory, (PRL), Ahmedabad in 1976. For the first two years observations were done using PbS photometer. Subsequently a liquid nitrogen cooled InSb photometer is in regular use. The astronomical observations are done from the two existing observatory sites in the country. In the present paper the observed atmospheric extinction values are given and also the details of transformation from the instrumental system to a standard photometric system are discussed.

II. Observations

The observations were obtained with an InSb photometer developed at PRL, Ahmedabad using 1 m telescope at the two observatory sites Nainital and Kavalur. The geographical details of the observatories are given in Table 1.

Table 1
Observing site details

	Nainital	Kavalur
Latitude	29° 21'.6 N	12° 34'.6 N
Longitude	79° 27'.4 E	78° 49'.7 E
Altitude	1950 m	725 m

The InSb photometer is cooled to liquid nitrogen temperature and contains a set of 5 filters closely resembling the standard JHKLM photometric bands. The observations were, however, done mainly in JHK bands as these telescopes are basically designed for optical work resulting in large thermal background at L & M bands. A focal plane chopper operating near 15 HZ was used to reject the background radiation from the sky. The filter characteristics are listed in Table 2 while the transmission characteristics are displayed in Figure 1.

Table 2
Interference filter characteristics

Photometric Band	Effective wavelength λ_e (μm)	Effective bandwidth $\Delta\lambda_e$ (μm)	Percentage Transmission
J	1.26	0.27	69
H	1.65	0.28	77
K	2.23	0.39	73

A fixed diaphragm of 20 arcsec in diameter was used throughout; the data used in the present analysis were obtained in 10 observing runs between November 1978 and December 1983.

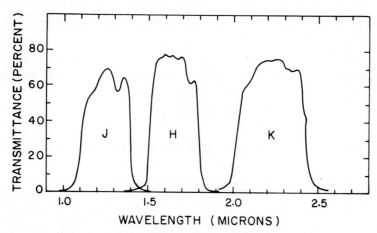

Fig. 1 : Transmittance curves of JHK filters.

III. Results

The atmospheric extinction coefficients were calculated using the linear law. The air-mass values needed for this calculation were estimated by the formula given by Hardie (1962); however, as most of the observations were limited to zenith angles upto 60° the air-mass is well represented by sec(z) term only. The difference in the average extinction values at Nainital and Kavalur was within the observational errors. The mean JHK extinction coefficients are listed in Table 3.

Table 3

Mean JHK extinction coefficients

k_J	k_H	k_K
0.15 ± 0.02	0.07 ± 0.02	0.09 ± 0.02

The magnitudes and colours in the local system were transformed to the standard system using the following equations:

$$(J-K) = \mu_{JK} (j-k)_o + \zeta_{JK}$$

$$(H-K) = \mu_{HK} (h-k)_o + \zeta_{HK}$$

$$K = k_o + \varepsilon_K (J-K) + \zeta_K$$

All the observations have been transformed into the standard photometric system defined by Johnson (965). For this purpose we observed standard stars from the list of Johnson et al. (1966) covering a reasonable range in K magnitudes and (J-K) & (H-K) colours. The transformation coefficients are listed in Table 4 and the colour transformations are graphically displayed in Figure 2a and b.

IV. Discussion

Of the two main parameters that determine the atmospheric extinction coefficients one is the filter transmission characteristics and other is the amount of precipitable water vapour in the atmosphere which depends on the observing site. The transmission characteristics of the interference filters employed in PRL photometer closely resemble the ones used in Kitt Peak National Observatory equipment (Manduca & Bell 1979). The observed values of precipitable water vapour content at Nainital during the winter observing season from November

Table 4

Mean transformation coefficients

Date	Month	Year	Place	μ_{JK}	ζ_{JK}	μ_{HK}	ζ_{HK}	ϵ_K	ζ_K
16-17	Feb	1980	Kavalur	1.107	0.72	1.076	0.45	-0.155	8.22
18-19	Feb	1980	Kavalur	0.934	0.70	0.866	0.49	-0.130	8.66
23-24	Apr	1980	Nainital	1.036	0.75	0.857	0.43	+0.090	8.93
15-16	Oct	1980	Nainital	0.984	0.70	0.757	0.33	+0.125	8.72
11-12	Jan	1981	Kavalur	0.992	0.69	0.551	0.26	+0.060	8.93
15-16	Feb$	1981	Kavalur	1.056	0.85	0.563	0.29	+0.004	7.51
16-17	Mar	1981	Nainital	1.071	0.76	1.053	0.41	+0.084	9.08

$ observations done with 20" telescope.

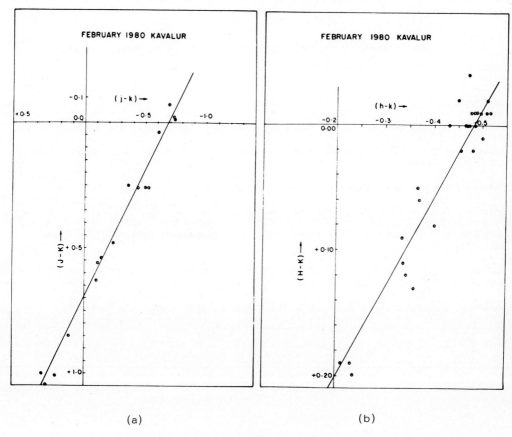

Fig.2a & b. Transformation curves.

to March lie in the range 2.5 to 4 mm (Kulkarni et al. 1977; Bhatt & Mahra 1987). Similar measurements at Kavalur are not available. However, because of lower altitude compared to Nainital and high relative humidity, the precipitable water vapour content at Kavalur is expected to be more than that at Nainital. Manduca & Bell (1979) have shown that the extinction between 1 and 2 air masses increases very slowly with increasing precipitable water vapour content. For an increase from 1.7 to 10.1 mm of precipitable water vapour the extinction coefficient increases by 0.04 mag at J and 0.02 mag. at K. As our observational errors lie in this range the average extinction values at Nainital and Kavalur are similar and in general agreement with the theoretical extinction coefficients calculated by Manduca & Bell (1979) for 5 mm precipitable water vapour and the Kitt Peak JHK filters.

Acknowledgements:

The research work at Physical Research Laboratory is funded by the Department of Space, Government of India.

References:

Bhatt, B.C. and Mahra, H.S. : 1987, Bull. Astron. Soc. India, **15**, 116.

Hardie, R.H. : 1962 in Astronomical Techniques, ed. W.A.Hiltner (Chicago: Univ. of Chicago Press), p.179.

Johnson, H.L. : 1965, Comm.Lun.Plan.Lab., **3**, 73.

Johnson, H.L. : 1966, Ann. Rev. Astron. Astrophys., **4**, 99.

Johnson, H.L., Mitchel, R.I., Iriarte, B. and Wisniewski, W.A. : 1966, Comm. Lun. Plan. Lab., **4**, 99.

Kulkarni, P.V., Ananth, A.G., Sinvhal, S.D. and Joshi, S.C. : 1977, Bull. Astron. Soc. India, **5**, 52.

Manduca, A. and Bell, R.A. : 1979, Pub. Astron. Soc. Pac., **91**, 848.

REDUCING PHOTOMETRY BY COMPUTING ATMOSPHERIC TRANSMISSION

Robert L. Kurucz
Harvard-Smithsonian Center for Astrophysics
60 Garden Street, Cambridge, MA 02138, USA

ABSTRACT

The transmission spectrum of the earth's atmosphere at every observatory can be computed if monitors are set up to determine the atmospheric structure and the abundance versus height of components that vary, such as water vapor and particulates. Photometric observations can be modelled and reduced using the measured instrumental bandpasses and the computed transmission. This method of reduction will greatly improve the quality of infrared photometry and may even be relevant in the visible where ozone, oxygen dimer, and water vapor affect photometric bandpasses. Here I describe the beginning of my work on this approach.

Atmospheric transmission is an important factor in most observations taken from the ground, both photometric and spectroscopic. In photometry almost all effective bandpasses are affected by atmospheric lines or rapidly varying continuous or pseudo-continuous absorption. I have always found it remarkable that astronomers pay it so little attention even though it is measurable and computable. The Air Force Geophysics Laboratory has worked on atmospheric transmisson for years and produced line lists and programs for low and high resolution computation of transmission (Rothman et al. 1987; Clough et al. 1986; Kneizys et al. 1983). The Air Force data and programs are continually being improved.

I have developed sophisticated programs for computing spectra (Kurucz and Avrett 1981). Given a solar, stellar, or terrestrial model atmosphere and a line list I can compute a 500000 point spectrum in one run. The lines have Voigt profiles with radiative, Stark, and van der Waals damping. A depth-dependent or constant microturbulent velocity field and a depth-dependent or constant Doppler shift can be specified. Once the opacity has been computed, a flux, intensity, or transmission spectrum can be determined and further broadened by macro-turbulence and by an instrumental profile. The effects of transmitting a solar or stellar spectrum through the earth's atmosphere followed by instrumental broadening can be computed. The computed spectra are plotted together with an observed spectrum and intercompared to determine the properties of the sun or star, or of the atmosphere, or both.

The line lists consist of data for over 48 million atomic and diatomic molecular lines appropriate for the sun and stars plus the AFGL atmospheric line list. I am working to improve and extend both the high and low temperature line data. I expect to have to make improvements in my treatment of broadening.

I am currently reducing Kitt Peak FTS spectra taken by James Brault in the visible and infrared for solar flux, solar intensity, and for high airmass sunsets. A solar flux atlas has already been published for 296 to 1300 nm (Kurucz, Furelid, Brault, and Testerman 1984). An infrared central intensity atlas from 1000 to 5400 nm has been published by Delbouille, Roland, Brault, and Testerman (1981). I plan eventually to publish atlases with all the solar and terrestrial lines labelled. Since the FTS scans have broad wavelength coverage, they should be redundant for determining atmospheric properties provided the input line data are accurate. The detailed atmospheric modelling is just getting under way. I expect to compute both the solar and terrestrial spectra and to iterate until I am sure of the transmission over as much of the spectrum as possible. Then I will divide it out to determine the spectrum above the atmosphere.

Both high resolution and high signal-to-noise are required for this work. Calculations are typically done at a resolution of 2000000. The observations typically have resolution of 500000 and signal-to-noise over 3000. I would like to have resolution of 1000000 and signal-to-noise of 10000 for widely varying atmospheric conditions to help resolve blends and to check damping treatments.

I am trying to fit the solar continuum in the visible and near infrared at the present time. I am having difficulties because there are broad structures produced by the Chappuis bands of ozone and by molecular oxygen "dimer" (Blickensderfer and Ewing 1969ab; Long and Ewing 1973) as shown in Figures 1 and 2. The oxygen "dimer" is not

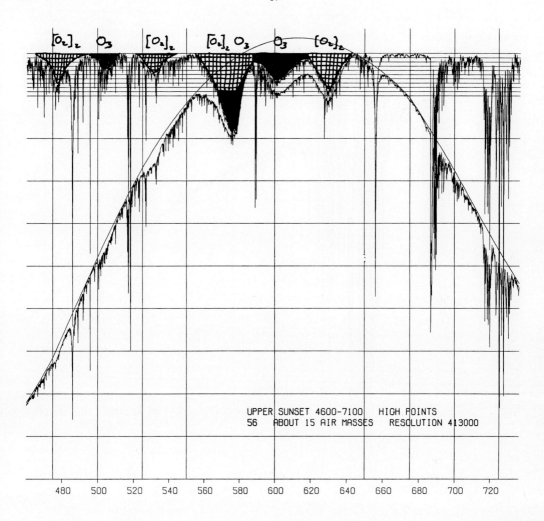

FIGURE 1. A sunset spectrum taken by James Brault at Kitt Peak through approximately 15 airmasses that shows broad absorption features of ozone and oxygen "dimer". The ozone features are just substructure in the much broader Chappuis bands. High points in the FTS spectrum centered at 600 nm are plotted and fitted with a "continuum" envelope. The high points are replotted at the top normalized to this "continuum". Even at 1 airmass these features are still significant.

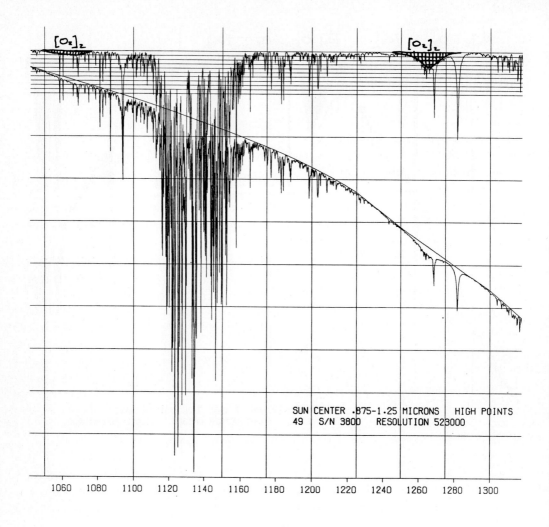

FIGURE 2. A 1 airmass central intensity spectrum as above showing the oxygen "dimer" features at 1060 and 1260 nm.

treated by the AFGL programs so they cannot reproduce this behavior, and its strength varies with the square of the density so it is very height dependent. It produces about a 1% effect in any observation from Kitt Peak at these wavelengths.

In the course of this work I must necessarily develop the ability to compute the transmission spectrum accurately. I expect to be able to find lines, line ratios, line profiles, and spaces between lines, that are sensitive diagnostics of atmospheric structure and abundances, even at relatively low resolution. Then cheap, mass-produced spectrometers could be designed (by someone else) and set up on an automatic telescope at every observatory (by someone else) to monitor the atmosphere throughout the night (and day).

The absorption spectrum of each component of the atmosphere can be pretabulated once and for all as a function of temperature and pressure. These spectra can be stored on a laser disk and simply interpolated to model the whole atmosphere. Once the temperature structure, water vapor, etc., have been determined, it will be possible to generate the whole transmission spectrum in a few minutes. The reduction time should not be significantly longer in spite of the increased complexity.

REFERENCES

Blickensderfer, R.P. and Ewing, G.E. 1969ab.
    Collision-induced absorption spectrum of gaseous oxygen at low temperatures and pressures. I. Journ. Chem. Phys., vol. 51, pp. 873-883; and, II. pp. 5284-5289.
Clough, S.A., Kneizys, F.X., Shettle, E.P., and Anderson, G.P. 1986.
    Atmospheric radiance and transmittance: FASCOD2, in Proceedings, Sixth Conference on Atmospheric Radiation, Williamsburg, VA.
Delbouille, L., Roland, G., Brault, J., and Testerman, L. 1981.
    Photometric Atlas of the Solar Spectrum from 1,850 to 10,000 cm^{-1}. Kitt Peak National Observatory, Tucson, 177pp.
Kneizys, F.X., Shettle, E.P., Gallery, W.O., Chetwynd, J.H.Jr., Abreu, L.W., Selby, J.E.A., Clough, S.A., and Fenn, R.W. 1983.
    Atmospheric transmittance/radiance: computer code LOWTRAN 6, AFGL-TR-83-0187 and supplement.
Kurucz, R.L. and Avrett, E.H. 1981.
    Solar spectrum synthesis. I. A sample atlas from 224 to 300 nm. Smithsonian Astrophys. Obs. Spec. Rep. No. 391, 139 pp.

Kurucz, R.L., Furenlid, I., Brault, J., and Testerman, L. 1984.
   <u>Solar Flux Atlas from 296 to 1300nm</u>. National Solar Observatory,
   Sunspot, New Mexico, 240 pp.
Long, C.A. and Ewing, G.E. 1973.
   Spectroscopic investigations of van der Waals molecules. I. The
   infrared and visible spectra of $(O_2)_2$. Journ. Chem. Phys., vol.
   58, pp. 4824-4834.
Rothman, L.S., Gamache, R.R., Goldman, A., Brown, L.R., Toth, R.A.,
   Pickett, H.M., Poynter, R.L., Flaud, J.-M., Camry-Peyret, C.,
   Barbe, A., Husson, N., Rinsland, C.P., and Smith, M.A.H. 1987.
   The HITRAN database: 1986 edition. Applied Optics, vol. 26,
   pp. 4058-4097.

DISCUSSION

Q.(A.T. Young): How do your high resolution models compare to HITRAN?

A: Right now I am using their line data (AFGL HITRAN line tape).
The spectrum is similar. The main difference is that my programs are
designed for large scale computing, the whole spectrum at once, but
they do a small region at a time. Once I have completely reduced the
atlases I expect to have corrected intensities and positions for some
of their lines.

JHKLM PHOTOMETRY:
STANDARD SYSTEMS, PASSBANDS AND INTRINSIC COLORS

M. S. Bessell and J. M. Brett
Mount Stromlo and Siding Spring Observatory
The Australian National University
Canberra A.C.T. 2611 Australia

Abstract

The relations between colors of the JHKL systems of SAAO, ESO, CIT/CTIO, MSO, AAO and Arizona have been examined and linear relations derived to enable transformation between the J-K, J-H, H-K, and K-L colors in the different systems. A homogenized system, essentially the Johnson-Glass system is proposed and its absolute calibration derived based on the Bell model atmosphere fluxes for Alpha Lyrae. The homogenized colors of the standard stars were used to derive intrinsic colors for stars with spectral types between B7V and M6V, and G7III and M5III. The JHKL passbands of the MSO IR system, derived from measured filter passbands and estimated atmospheric transmission values, were used to compute synthetic colors from relative absolute fluxes of some stars (including the sun). The reasonable agreement with the standardized JHKL colors indicates that these passbands can be adopted as representing the homogeneous system, and used to compute broad band IR colors from theoretical or observed fluxes. The passbands of other IR systems were similarly estimated from published data, and the synthetic colors were intercompared using black-body and stellar fluxes. These passbands were then adjusted in wavelength to produce agreement with the observed relations between different systems, enabling the effective wavelengths of the different natural systems to be established. Better effective wavelengths could be determined were spectrophotometry available for the very red stars with known broad band colors. The full text of this paper is published in Bessell and Brett (1988).

Summary

Many natural JHKL photometric systems exist and although they have common ancestory in the JKLM system of Johnson, Mitchell, Iriarte, and Wisniewski (1966) and the JHKL system of Glass (1974), and in their InSb detectors, the filters are not identical. Standardization has mainly involved adoption of zero-points based on zero colors for an "A0" star, and not adjustment for any differences in the filter effective wavelengths. The J passbands in particular are known to differ between the MSO, AAO, CIT/CTIO and Johnson systems. However, when we computed synthetic IR colors using measured filter passbands for the MSO system it was apparent that the J band was not at the effective wavelength inferred from published comparisons of the various systems. Arbitrarily shifting the passband to the red helped somewhat, but the shift was far greater than could be justified by any uncertainties, such as atmospheric cutoff. Similar difficulties were encountered when computing colors for published passbands of other systems. Clearly we needed to reexamine the basis of passband and effective wavelength derivations. Since Koornneef's analysis (Koornneef, 1983) of the magnitudes of the SAAO, ESO, AAO and Johnson systems, we now have a more extensive list of MSO standards (Hyland and McGregor, 1986) containing many M dwarfs, the comparison of the AAO (Allen and Cragg, 1983) and CIT/CTIO (Elias et al, 1982) systems (Elias, Frogel, Hyland and Jones 1983), some L' and M standards of MKO (Sinton and Tittemore, 1984) and the important review of Glass (1985), to supplement the data available to Koornneef. There is also extensive IR photometry of M dwarfs by Stauffer, and Hartman (1986), and photometry of Hyades dwarfs by Carney (1982). In this paper, we discuss the intercomparisons between colors measured in different systems for common stars, and also examine the mean relations between "pure" IR colors (in the different systems) such as J-K, H-K, and K-L and the "mixed" color V-K and the near-IR color V-I of the Kron-Cousins system. This enables us to use measurements for more stars than those in common, and derive more reliable zero-point corrections. Using the results of the comparisons, a homogenized system was produced and intrinsic colors derived for different spectral types.

Finally, we discuss the passbands of the MSO system, for which we have the most information, and compare these with passbands of the other major systems, by intercomparing black-body colors and colors from observed spectra of some KM giants. In most cases it was necessary to shift the passbands in effective wavelength to produce near agreement with the observed linear relations between the different system colors and those of the MSO system. The adopted MSO passbands appear to reasonably reproduce the "standard", homogenized colors, and so can be used to compute broad-band colors from IR spectra or model atmosphere fluxes.

Table 1   Adopted Linear Transformation Equation Coefficients

	SAAO	Johnson	ESO	MSO	AAO	CIT	Carter	HCO	S&H	Carney
J-H	-0.005	0.01	-0.010	0.01	-0.002	-0.025	0.00	0.00	0.06	0.02
	1.00	1.01	1.105	1.016	0.963	1.098	0.94	1.01	0.97	1.09
H-K	-0.021	0.01:	0.005	-0.007	-0.003	0.00	0.004	0.02	0.005	0.005
	1.0	0.91:	0.87:	0.97	0.98	1.03	0.994	1.0	1.03	1.0
J-K	-0.005	0.01	-0.01	0.01	0.0	-0.002	0.02	0.05	0.026	
	1.0	0.99	1.025	1.008	0.974	1.086	0.975	1.0	1.0	1.0
K-L	0.0	-0.03	-0.03		*	0.0	0.01			
	1.0	1.0	1.0			1.0	0.80:			
V-K	-0.005	0.01	0.015	-0.012	0.0	-0.02	0.005	0.0:	0.0:	0.0:
	1.0	0.993	1.0	0.997	1.002	1.001	1.0	1.0:	1.0:	1.0

* AAO has a 3.8 micron filter which defines the K-L' color; MKO has a similar filter and K-L'=1.04(K-L')MKO

Table 2.   Intrinsic Colors for Dwarfs

MK	V-I	V-K	J-H	H-K	J-K	K-L	K-L'	K-M
B8	-0.15	-0.35	-0.05	-0.035	-0.09	-0.03	-0.04	-0.05
A0	0.00	0.00	0.00	0.00	0.00	0.00	0.00	0.00
A2	0.06	0.14	0.02	0.005	0.02	0.01	0.01	0.01
A5	0.27	0.38	0.06	0.015	0.08	0.02	0.02	0.03
A7	0.24	0.50	0.08	0.02	0.10	0.03	0.03	0.03
F0	0.33	0.70	0.12	0.025	0.15	0.03	0.03	0.03
F2	0.40	0.82	0.15	0.03	0.18	0.03	0.03	0.03
F5	0.53	1.10	0.22	0.035	0.25	0.04	0.04	0.02
F7	0.62	1.32	0.27	0.04	0.32	0.04	0.04	0.02

MK	V-I	V-K	J-H	H-K	J-K	K-L	K-L'	K-M
G0	0.66	1.41	0.29	0.05	0.36	0.05	0.05	0.01
G2	0.68	1.46	0.30	0.052	0.37	0.05	0.05	0.01
G4	0.71	1.53	0.32	0.055	0.385	0.05	0.05	0.01
G6	0.75	1.64	0.35	0.06	0.43	0.05	0.05	0.00
K0	0.88	1.96	0.42	0.075	0.53	0.06	0.06	-0.01
K2	0.98	2.22	0.48	0.09	0.59	0.07	0.07	-0.02
K4	1.15	2.63	0.58	0.105	0.68	0.09	0.10	-0.04
K5	1.22	2.85	0.62	0.11	0.72	0.10	0.11	
K7	1.45	3.16	0.66	0.13	0.79	0.11	0.13	
M0	1.80	3.65	0.695	0.165	0.86	0.14	0.17	
M1	1.96	3.87	0.68	0.20	0.87	0.15	0.21	
M2	2.14	4.11	0.665	0.21	0.87	0.16	0.23	
M3	2.47	4.65	0.62	0.25	0.87	0.20	0.32	
M4	2.86	4.98	0.60	0.275	0.88	0.23	0.37	
M5	3.39	5.84	0.62	0.32	0.94	0.29	0.42	
M6	4.18	7.30	0.66	0.37	1.03	0.36	(0.48)	

Table 3   Intrinsic Colors for Giants (class III)

MK	V-I	V-K	J-H	H-K	J-K	K-L	K-L'	K-M
G0	0.81	1.75	0.37	0.065	0.45	0.04	0.05	0.00
G4	0.91	2.05	0.47	0.08	0.55	0.05	0.06	-0.01
G6	0.94	2.15	0.50	0.085	0.58	0.06	0.07	-0.02
G8	0.94	2.16	0.50	0.085	0.58	0.06	0.07	-0.02
K0	1.00	2.31	0.54	0.09	0.63	0.07	0.08	-0.03
K1	1.08	2.50	0.58	0.10	0.68	0.08	0.09	-0.04
K2	1.17	2.70	0.63	0.115	0.74	0.09	0.10	-0.05
K3	1.36	3.00	0.68	0.13	0.82	0.10	0.12	-0.06
K4	1.50	3.26	0.73	0.14	0.88	0.11	0.14	-0.07
K5	1.63	3.60	0.79	0.165	0.95	0.12	0.16	-0.08
M0	1.78	3.85	0.83	0.19	1.01	0.12	0.17	-0.09
M1	1.90	4.05	0.85	0.205	1.05	0.13	(0.17)	-0.10
M2	2.05	4.30	0.87	0.215	1.08	0.15	(0.19)	-0.12
M3	2.25	4.64	0.90	0.235	1.13	0.17	(0.20)	-0.13
M4	2.55	5.10	0.93	0.245	1.17	0.18	(0.21)	-0.14
M5	3.05	5.96	0.95	0.285	1.23	(0.20)	(0.22)	-0.15
M6		6.84	0.96	0.30	1.26			0.0:
M7		7.8	0.96	0.31	1.27			0.0:

Table 4   Adopted Passbands

J		H		K		L		L'		M	
1040	0.00	1460	0.00	1940	0.00	3040	0.00	3440	0.00	4440	0.00
1060	0.02	1480	0.15	1960	0.12	3080	0.02	3480	0.02	4480	0.13
1080	0.11	1500	0.44	1980	0.20	3120	0.09	3520	0.19	4520	0.34
1100	0.42	1520	0.86	2000	0.30	3160	0.38	3560	0.80	4560	0.30
1120	0.32	1540	0.94	2020	0.55	3200	0.30	3600	0.90	4600	0.39
1140	0.47	1560	0.98	2040	0.74	3240	0.50	3640	0.91	4640	0.50
1160	0.63	1580	0.95	2060	0.55	3280	0.61	3680	0.85	4680	0.44
1180	0.73	1600	0.99	2080	0.77	3320	0.41	3720	0.82	4720	0.16
1200	0.77	1620	0.99	2100	0.85	3360	0.50	3760	0.86	4760	0.35
1220	0.81	1640	0.99	2120	0.90	3400	0.61	3800	0.88	4800	0.33
1240	0.83	1660	0.99	2140	0.94	3440	0.70	3840	0.86	4840	0.37
1260	0.88	1680	0.99	2160	0.94	3480	0.85	3880	0.91	4880	0.44
1280	0.94	1700	0.99	2180	0.95	3520	0.88	3920	0.97	4840	0.37

J		H		K		L		L'		M	
1300	0.91	1720	0.95	2200	0.94	3560	0.84	3960	0.97	4880	0.44
1320	0.79	1740	0.87	2220	0.96	3600	0.84	4000	0.97	4920	0.44
1340	0.68	1760	0.84	2240	0.98	3640	0.84	4040	0.76	4960	0.37
1360	0.04	1780	0.71	2260	0.97	3680	0.86	4080	0.63	5000	0.37
1380	0.11	1800	0.52	2280	0.96	3720	0.65	4120	0.54	5040	0.23
1400	0.07	1820	0.02	2300	0.91	3760	0.19	4160	0.24	5080	0.07
1420	0.03	1840	0.00	2320	0.88	3800	0.04	4200	0.00	5120	0.00
1440	0.00			2340	0.84	3840	0.00				
				2360	0.82						
				2380	0.75						
				2400	0.64						
				2420	0.21						
				2440	0.10						
				2460	0.03						
				2480	0.00						

Table 5   Effective Wavelengths[1], Zeropoint Fluxes[2] and Magnitudes[3]

	V	J	H	K	L	L'	M	(M)
λ_{eff}	0.545	1.22	1.63	2.19	3.45	3.80	4.75	4.80
ZP	0.000	0.90	1.37	1.88	2.77	2.97	3.42	3.44
F_λ	3590	312	114	39.4	6.99	4.83	2.04	1.97
F_ν	3600	1570	1020	636	281	235	154	152

[1] In μm
[2] F_λ (10^{-15} W cm^{-2} μm^{-1}), F_ν (10^{-30} W cm^{-2} hz^{-1}) for a 0.03 magnitude star from Dreiling and Bell and Bell Vega models for adopted passbands.
[3] Mag = $-2.5 \log<F_\nu> - 66.08 - ZP$

References

Allen, D. A., and Cragg, T. A. 1983, Mon. Not. R. Astron. Soc. 203, 777.
Bessell, M. S., and Brett, J. M. 1988, Pub. Astron. Soc. Pacific 100, 1134.
Carney, B. W., 1982, Astron. J. 87, 1527.
Dreiling, L. A., and Bell, R. A. 1980, Astrophys. J. 241, 736.
Elias, J. H., Frogel, J. A., Matthews, K., and Neugebauer, G., 1982, Astron. J. 87, 1029.
Elias, J. H., Frogel, J. A., Hyland, A. R. and Jones, T. J. 1983, Astron. J. 88, 1027.
Glass, I. S. 1974, MNASSA 33, 53 and 71.

Glass, I. S. 1985, Irish Astron. J. 17, 1.

Hyland, A. R., and McGregor, P. J. 1986, private communication.

Johnson, H. L., Mitchell, R. I., Iriarte, B., and Wisniewski, W. Z. 1966, Comm. Lunar and Planetary Lab. 4, 99.

Koornneef, J. 1983, Astron. Astrophys. 128, 84.

Sinton, W. M., and Tittemore, W. C. 1984, Astron. J. 89, 1366.

Stauffer, J. R., and Hartman, L. W. 1986, Astrophys. J. Suppl. 61, 531.

STANDARDIZATION WITH INFRARED ARRAY PHOTOMETERS

Ian S. McLean
Joint Astronomy Centre, Hilo, Hawaii
and Royal Observatory, Edinburgh, United Kingdom

Abstract

A brief introduction is given concerning the recent and rapid increase in the use of solid-state infrared imaging devices for photometry at infrared wavelengths. Parallels are drawn with optical CCD photometry and some of the problem areas are discussed.

1. INTRODUCTION

Over the last two years infrared array detectors analogous to CCDs at optical wavelengths, having become available at several major international facilities, are now used for photometric imaging to very faint levels. The first of these array cameras to come into operation as a common-user or "facility" instrument was the IRCAM on the 3.8m UK Infrared Telescope on Mauna Kea (McLean et al. 1986, McLean 1987). That camera employs a 62x58 element array of InSb detectors from SBRC (U.S.A.). Similar systems are now in operation at Kitt Peak and Cerro Tololo, as well as at several other observatories or in travelling systems. At the NASA IRTF, Steward Observatory, University of Chicago and at ESO, another form of infrared array sensor has been used, namely, HgCdTe (Mer-cad-tel). Unlike InSb which operates over the wavelength range 1.0 to 5.1 µm, HgCdTe arrays for astronomy applications are tailored to have a long wavelength cut-off between 2.5 and 3.5µm. Array formats of 64x64 have been available for over 2 years but 128x128 devices have now appeared. InSb arrays must be cooled to 35K (approximately) to achieve the same dark current (~100 e/s) as HgCdTe arrays do at 77K. Also, a 256x256 array of PtSi Schottky Barrier diodes has been used. These devices are very uniform but they have low quantum efficiency (~ a few percent) and a strong wavelength dependence of response; they are not normally sensitive beyond 2.2µm. For longer wavelengths, small arrays of doped silicon (e.g., Si:Ga) are now in use at 10µm.

The significance of these developments cannot be underestimated. For instance, IRCAM receives over 60% of the telescope time on UKIRT and many astronomers are choosing to do photometry with the new array cameras instead of single-aperture photometers. The new cameras are extremely sensitive, as good or better per pixel as the best single-channel detectors, and are ideal for surface

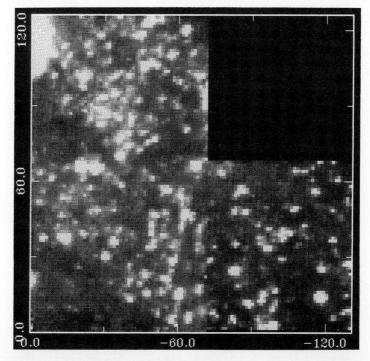

Figure 1. A region near the Galactic Center (upper left) at 1.65μm.

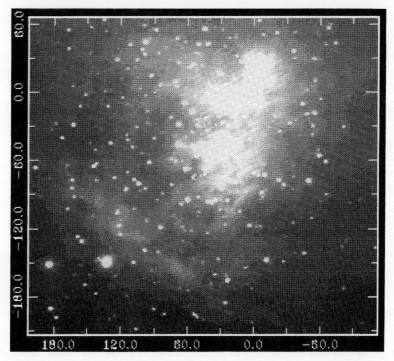

Figure 2. The Orion nebula at 2.2μm with the UKIRT IRCAM.

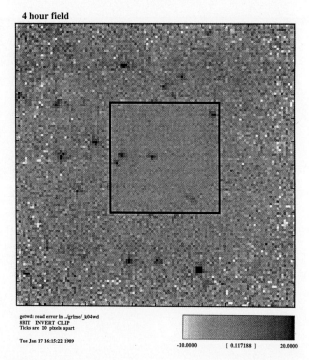

Figure 3. An infrared image of a field at high galactic latitude. The faintest objects have an integrated K magnitude of ~21.

photometry, faint object photometry and photometry in crowded fields such as star forming regions, the galactic centre and globular clusters.

A few representative examples of what has been achieved to date are shown in Figs. 1 to 3. Colour renditions of many results may be found in McLean (1989a, 1988b), and the proceedings of the Hawaii workshop on "Astronomy with Infrared Arrays" edited by Wynn-Williams and Becklin (1987) contains many reviews.

2. PHOTOMETRY WITH IR ARRAYS

Just as with optical CCDs, infrared photometry from arrays must be performed "after-the-fact" by software manipulations on image frames. Procedures similar to those at optical wavelengths must be followed such as (dark + bias) subtraction, flat-fielding, linearity corrections, removal of bad pixels, software aperture photometry or point spread function (PSF) "fitting" photometry. There are a number of problems to be considered in addition to those of infrared atmospheric extinction and calibration of the array detector.

2.1 Passbands and wavelength response.

Most classical infrared photometers use indium antimonide detectors and so for the InSb arrays the wavelength response is very nearly the same. This is not

true for HgCdTe and PtSi arrays. But even for InSb there may be subtle
differences, due for example to anti-reflection coatings on the detector or to the
lower than normal detector temperature of 35K, which make small colour terms
inevitable in intercomparing systems. Also, the set of JHKL filters in common use
are not, in fact, identical. Different manufacturing runs and even different
manufacturers have been used over the years, and additional blocking filters have
been employed to eliminate red leaks (at J and H in particular). This problem, of
course, is shared with aperture photometry.

2.2 Faint standards.

The set of standards established by Elias for instance, are in the magnitude
range 3-7 at K (2.2μm) and were done with InSb single-aperture photometers. When
array cameras are used on large telescopes very faint limiting magnitudes are
possible. For example, recent results on UKIRT with IRCAM (Cowie et al. 1988)
have gone down to 1 sigma detection levels of K = 22 in 1."2 pixels; the K
background is 12.5 - 13.0 per sq. arcsec. Since the InSb array detectors are used
in reverse-biassed mode to create a storage capacity of one million electrons they
inevitably suffer some non-linearity due to the change in capacitance with
decreasing reverse bias. It is therefore important to calibrate this effect
(typically of order 8% in our system), which can be done from a series of flat
fields, and to check carefully against fainter point-sources of known magnitude.
The Elias stars are too bright for this purpose. We have had good success with
the technique of de-focussing the 7-magnitude standards, until their surface
brightness is less than K = 12.5, and then cross-checking against white dwarfs and
selected stars in globular clusters in the magnitude range 11 - 14 at K. However,
what is really required is a good set or sequence of equatorial standards down to
about K = 15.

2.3 Software photometry.

The issues here are very similar to those encountered by CCD photometrists.
After detector calibrations, which includes linearity corrections which may be a
function of light level in a pixel, photometric parameters must be derived by
analysis of the final image by a software algorithm. The simplest approach is
"aperture" photometry, usually done with two concentric apertures on the object.
Alternatively, point spread function (PSF) "fitting" routines can be used. Among
the software packages available for this purpose are DAOPHOT (IRAF), ROMOFOT
(MIDAS) and STARMAN (STARLINK).

Two major problems arise from the small size of the currently available array
detectors.
    (1)   undersampling
    (ii)   registration

Often, to maximize total field of view, the infrared image is undersampled at 0".6 to 1".0 per pixel, instead of say 0".3 to 0".5 per pixel. Even so, the total field for a 64x64 array is still small which leads to the use of "mosaics" and hence the need for software registration. Errors of registration inevitably creep in. Seeing variations during the time taken to make the mosaic mean that the PSF is not constant across the final image. Photometry should therefore be done on the individual frames rather than on the mosaic.

In the galactic plane and in starforming regions infrared cameras reveal a huge number of hidden sources. Likewise, deep cosmological imaging results in very crowded fields. These are regimes where cameras excel. Under such conditions software aperture photometry is unreliable and PSF techniques are needed, but their success depends on developing good algorithms to deal with under-sampled images.

3. Summary and conclusions.

Infrared photometry is now being done with cameras using infrared array detectors similar in many ways to optical CCDs. These cameras are much better suited to the study of star forming regions, the galactic centre, extended infrared sources, globular clusters, galaxies and deep cosmological surveys than previous single-aperture photometers. Many of the issues relating to extinction and standardization at infrared wavelengths apply equally well to array photometers as to single-aperture photometers. Additional problems are those of passband definition (less so for the InSb arrays), linearity, lack of faint standards, and serious undersampling of the images for software techniques relying on PSF fitting to derive magnitudes.

Some of these problems may be transitory and will be alleviated when IR arrays become as large as optical arrays, others will not. It is clear that if IR cameras are to be used for photometry as well as morphology then great attention to detail during the reduction process is essential, and in common with aperture-photometers, more effort on standardization is required.

References

Cowie, L., Lilly, S.J., Gardner, J., and McLean, I.S., 1988, "A Cosmologically Significant Population of Galaxies Dominated by Very Young Star Formation", Ap.J., Letters, $\underline{332}$, L29-32.

McCaughrean, M.J. 1988, Ph.D. Thesis, University of Edinburgh.

McLean, I.S., Chuter, T.C., McCaughrean, M.J., and Rayner, J.T., 1986, SPIE Vol. 627, 430.

McLean, I.S. 1987 in "Infrared Astronomy with Arrays", eds. C.G. Wynn-Williams and E.E. Becklin, Institute for Astronomy, University of Hawaii.

McLean, I.S. 1988a, "Infrared Astronomy's New Image", Sky and Telescope, March Vol. 75, No. 3, 254.

McLean, I.S. 1988b, "Infrared Astronomy: A new beginning", Astronomy Now, July Vol. 2, No. 7, 25.

Wynn-Williams, C.G. and Becklin, E.E. 1987, "Infrared Astronomy with Arrays", Proc. of the Hilo Workshop, Institute for Astronomy, University of Hawaii, Honolulu.

A SUMMARY OF THE SESSION

R. A. Bell
Astronomy Program
University of Maryland
College Park, MD 20742

I must say that I find it rather flattering to be asked to give the summary talk in a field where I have never done any observing at all.

In giving a summary talk, it is customary to review or to describe the contributions of the various contributors. I'll maintain that hallowed tradition. However, there are a number of points at issue, points which are rather broader than those which were discussed here today.

As we all know, the extinction in the infra-red region of the spectrum (which I'll take as being 0.76 μm to 5.0 μm for today's talk) is very complex. (An example of expected extinction shown in Fig. 1.). At some wavelengths (e.g. 1.35 μm, 1.85 μm and 2.60 μm) the terrestrial atmosphere is expected to be virtually opaque, owing to very strong absorption by lines of the water molecule. At other wavelengths, e.g. 1.6 and 2.2 μm, the molecular absorption is quite small. Fortunately, we have an extensive compilation of data for these molecular lines (McClatchey et al. 1973) and this has led to several attempts to model the absorption. In addition to the molecular data, such modelling requires information on the abundances of the various absorbers i.e. H_2O, O_3, CO_2, N_2O, CO, CH_4 and O_2 as well as pressure and temperature values. This latter data is needed to describe the line broadening as well as the excitation of the molecules. Very often such data is found from models of the terrestrial atmosphere e.g. the standard midlatitude winter and summer models given by McClatchey et al. (1972).

This modelling approach was used by Traub and Stier (1976), who were interested in comparing the atmospheric transmission computed for Mauna Kea with that which would be expected for an aeroplane at various altitudes. Traub and Stier have very kindly given copies of their program to other people and it has been used in various studies e.g. the work by Manduca and Bell (1979), as well as for Fig. 1. (Note that the absorption around 4.2 microns starts too abruptly in Fig. 1, a number of O_3 lines appearing to be missing from the data file.)

At this point I note that there are a number of improvements which can be made. Firstly, there is the question of the adequacy and accuracy of the line list. This is something which can be checked using solar atlases e.g. that of Delbouille et al. (1981). Secondly, the Traub and Stier program employs a single-layer model of the terrestrial atmosphere. It is gratifying to see that Volk, Clark and Milone (1989) have carried out calculations using a multi-layer model and even more gratifying that these calculations indicate very little difference between the single-layer model and the multi-layer model.

The calculations make a number of predictions. Firstly, the extinction in the infrared pass bands J, H, K and L depends a great deal on the location and width of the pass band e.g. the extinction for the Johnson system is much greater than that for systems with narrower filters e.g. that of Glass (1973). Secondly, the extinction depends strongly on the abundance of water vapor, giving a marked seasonal effect for Kitt Peak, for example, and clearly indicating that night-to-night fluctuations can occur. Thirdly, the extinction is not linear with air mass. Fourthly, the extinction in J does depend a little on the colour of the star being observed i.e. the extinction for Vega will be a little less than that for Aldebaran. Finally, the extinction at Mauna Kea is predicted to be a little less than that of Kitt Peak in the winter.

Do the other calculations and observations bear this out?

Ashok (1989) and Glass and Carter (1989) have presented data for the extinction at their observatories. Ashok notes general agreement with the calculations of Manduca and Bell (1979) while Glass and Carter give a more detailed discussion, remarking that the extinction in K seems to have been underestimated. This is a little puzzling, although, following the work of Angione the underestimate may be due to the neglect of scattering by dust and aerosols. Fig. 1 appears to explain the much greater variation in J and L zero points observed by Glass and Carter - these variations are presumably caused by the absorption at 1.15 microns and 3.2 - 3.5 microns, respectively. These variations at L suggest that the use of L as the reference band in the infra red flux method of determining stellar temperatures (see, for example, Saxner and Hammarbäck 1985) should probably be avoided. The K band seems superior. Fig. 1 also probably explains the remarks made by Bessell and Brett (1989) about the difficulties of observing in the M band.

I found Angione's (1989) talk to be particularly stimulating and informative with numerous valuable comments. In particular, I refer to his comment that aerosols are in the stratosphere and consequently observing from Mauna Kea will not give as great an advantage as might be hoped. In addition, the comment that the scattering by dust and aerosols might have the same effect at 1 micron as at 10 μm is a sobering one.

Young (1989) gave an interesting account of ways in which the Manduca and Bell calculations could be used in practice while Kurucz (1989) gave us his vision of the future.

Bessell and Brett (1989) considered a number of problems, emphasizing the question of transformation equations between different systems and noting the importance of water vapor and TiO in the stellar spectra. Bessell also touched upon the vexing question of the absolute calibration of Bega in the infrared--see the discussion by Mountain et al. (1985)--by noting that the model calibrations of Dreiling and Bell (1980) match those of Labs and Neckel (1970). Bessell also noted that the solar flux in some wavelength regions is derived from solar models but that these models omit the effects of the solar CO lines.

Finally, McLean (1989) regaled us with some very beautiful pictures obtained with a camera which uses the Santa Barbara Research Center infrared array. I'd like to emphasize a point made by Kurucz on the use of this camera, namely that its great increase in sensitivity compared to other detectors offers at least the possibility of carrying out observations using filters which are much narrower than those used now. The use of narrower filters at suitable wavelengths would greatly alleviate the problems of extinction that were discussed here today.

I'd like to thank the speakers for making this such a very interesting session. Ms. Leslie Wisz kindly prepared Fig. 1 for me. My work on the spectra of cool stars is supported by the National Science Foundation under grant AST85-13872.

References

Angione, R. J., 1989, "Atmospheric Extinction in the Infrared", Proceedings of the Sessions on Problems of Infrared Extinction and Standardization (IAU Commissions 9 and 25), Baltimore, Maryland, August 4, 1988 (Berlin: Springer-Verlag), p. 26.

Ashok, N. M., 1989, "Near-Infrared Extinction Measurements at the Indian Observatory Sites", Proceedings of the Sessions on Problems of Infrared Extinction and Standardization (IAU Commissions 9 and 25), Baltimore, Maryland, August 4, 1988 (Berlin: Springer-Verlag), p. 50.

Bessell, M. S. and Brett, J. M., 1989, "JHKLM Photometry: Standard Systems", Passbands and Intrinsic Colors", Proceedings of the Sessions on Problems of Infrared Extinction and Standardization (IAU Commissions 9 and 25), Baltimore, Maryland, August 4, 1988 (Berlin: Springer-Verlag), p. 62.

Delbouille, L., Roland, G., Brault, J. and Testerman, L. 1981, "Photometric Atlas of the Solar Spectrum from 1850 to 1000 cm^{-1}", Kitt Peak National Observatory.

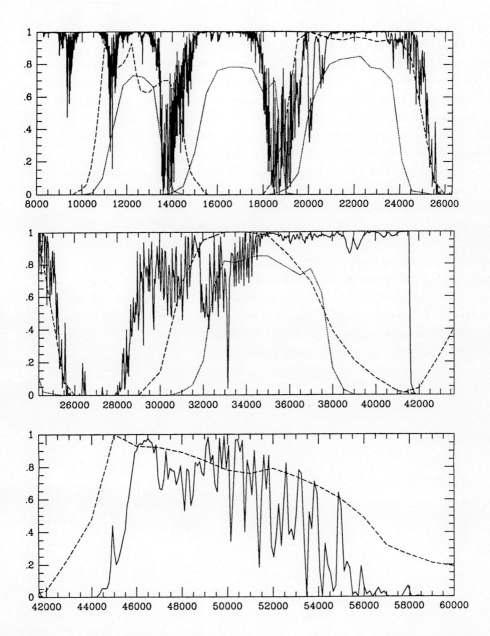

Figure 1. The transmission of the terrestrial atmosphere, computed for Kitt Peak winter conditions (Manduca and Bell 1979) is plotted versus wavelength (in Å). Also plotted are the sensitivity functions of the Johnson system (dashed lines) and Caltech - CTIO system (dotted lines).

Dreiling, L. A., and Bell, R. A., 1980, Ap. J., 241, 736.

Glass, I. S., 1973, M.N.R.A.S., 164, 155.

Glass, I. S. and Carter, B.S., 1989, "Infrared Extinction at Sutherland", Proceedings of the Sessions on Problems of Infrared Extinction and Standardization (IAU Commissions 9 and 25), Baltimore, Maryland, August 4, 1988 (Berlin: Springer-Verlag), p. 38.

Labs, D. and Neckel, H., 1970, Solar Phys., 15, 79.

Manduca, A. and Bell, R. A., 1979, P.A.S.P. 91, 848.

McClatchey, R. A., Benedict, W. S., Slough, S. A., Burch, D. E., Calfee, R. F., Fox, K., Rothman, L. S., and Garing, J. S., 1973, "AFCRL Atmospheric Absorption Line Parameter Compilation", AFCRL-TR-73-0096, Bedford, Mass.

Mountain, C. M., Leggett, S. K., Selby, M. J., Blackwell, D. E., and Petford, A. D., 1985, Astr. Astrophys. 151, 399.

Saxner, M. and Hammarbäck, G., 1985, Astr. Astrophys., 151, 372.

Traub, W. A., and Stier, M. T., 1976, Appl. Optics 15, 364.

Volk, K., Clark, T. A. and E. F. Milone, 1989, "Models of Infrared Atmospheric Extinction", Proceedings of the Sessions on Problems of Infrared Extinction and Standardization (IAU Commissions 9 and 25), Baltimore, Maryland, August 4, 1988, (Berlin: Springer-Verlag), p. 16.

CONCLUDING POSTSCRIPT

E. F. Milone
The Rothney Astrophysical Observatory
The University of Calgary
Calgary, AB   T2N 1N4/Canada

It is a useful prerogative to have the final 'word'. It permits me to add a few comments regarding some of the immediate practical consequences of the meeting, and it is in keeping with my suggestion that this record of the proceedings state what we had intended or wished we had said, rather than what we actually did say!

First, in the Introduction, I asked a number of questions, few of which were explicitly answered. However, the kinds of investigations undertaken by Glass and colleagues at Sutherland have the power to provide confidence that 'differential' infrared photometry, involving objects widely separated spatially and temporally, can be trusted. For purely differential near-IR work, I still argue for 'InfraRed Alternate Detection System' or IRRADS photometry which can, in principle, permit rapid sequential observations of target and comparison star with sky stations between and beyond them (cf. Milone et al. 1982, Milone and Robb 1983 for the basic idea). The operation of an infrared 'RADS' would be more complicated than its optical counterpart for a variety of reasons, but I am convinced that it is a viable technique. On the basis of the similarity of shape of infrared extinction curves, Young suggests that extinction is determinable from standard star observations if (only) 50% more time is devoted to extinction determination and if the air mass is below ~ 3 to minimize the effects of the variable scale height of water vapor. It is a woeful fact, however, that even optical extinction work is often not performed as thoroughly as one would like. Nevertheless, this approach should provide intermediate precision absolute photometry and hence respectable precision in differential photometry. Combining results from different nights does not strike me as an effective course of action because of the clearly variable water vapor content (to say nothing of dust and other constituents). I think that the modelers are correct when they insist that we need to know the water vapor content of the atmosphere as we are observing through it, if we truly want to know the extinction at any instant. It is not too much to ask of a modern observatory to have a computer perform dynamic modeling in the background, if the water vapor can be monitored in some automated way. Various water vapor meters have been devised but usually require the Sun or Moon as source; others use emission of the daytime sky. Perhaps a standardized system can be made with improved sensitivity to operate on bright stars. On the hopeful side, Angione described a clever technique which should be tried, and water vapor meter tests involving calibration against balloon-borne radiosonde measurements are being conducted at Mauna Kea. Bell noted in his summary that the extensive information required for the modeling process itself involves many assumptions about the instantaneous state of the atmosphere.

Second, Volk et al. noted that the nonlinear character of infrared extinction was recognized by Harold Johnson more than two decades ago, but that it is not clear whether or not Johnson's suggested solution - to extend the Bouguer linear extinction to a negative airmass - was actually carried out for any of Johnson's data. In an unrecorded comment during the sessions, W. Z. Wisniewski mentioned that until recently, at least, these data were available in Arizona. Hopefully, they will continue to be preserved so that further work can be done on them. The full, unpublished set of observations ought to be able to better define Johnson's system. The importance of maintaining such observational data bases cannot be overemphasized. The extensive infrared programs undertaken at the South African, Australian, and Indian observatories which were reported here underscore the value of such data.

Third, one of the frequently mentioned ways of improving the extinction and transformation process - by redesigning the filter passbands to be narrower and better centered in the atmospheric windows - is being explored actively at present writing. Bessell noted the advantages of using the L' filter, shifted further to the red than the Johnson L, which is in fact at the blue edge of the window (originally designed for use with the PbS detector which Johnson used). He and others also suggested that there appeared little value in the M filter on the grounds that the window scarcely exists, thus exacerbating the types of problems discussed in the present meeting, that the M region is blanketed by CO in all the cooler stars, that K-L' is a better temperature indicator for the coolest stars, and that photometry at 10 μm (i.e. at the N passband) suffices to indicate circumstellar emission. Subsequent to the meeting, however, a member of the SOC was informed that solar system observers rely on the M-band and that we should not recommend its abandonment. It would seem, however, that if stellar calibration is part of the process, some improvement in definition is appropriate. In response to a post-meeting inquiry about the availability of filters, J. A. Dobrowolski of the National Research Council of Canada indicated that filters of 50 - 100 nm width, with good blocking properties, good stability at LN_2 and LHe temperatures, and size range (for imaging as well as aperture photometry) are well within the capabilities of commercial filter manufacturers. In light of the difficulties to be faced were any existing system proclaimed as a preferred standard system, it seems reasonable to organize a broadly constituted working group to work on recommendations for the selection of suitable filter definitions. If these meet with general approval, implementation could follow.

Fourth, even if a common system fails to be established or accepted, we have seen here ample evidence of the value of properly defining one's system and making these details available to the community. In other words, infrared observers, know thy system by providing the rest of us with transformation coefficients and zero points – and do not ignore extinction on the way!

Finally, it is time to begin to pay attention to the problems described by Ian McLean – infrared imagers are already detectors of choice in the nearer IR. Further discussions of the capabilities and limitations of these imagers are highly desirable.

It is a pleasure to thank all the participants and the Scientific Organizing Committee, referees, and everyone else who participated directly and indirectly in helping to make the sessions successful, and this volume cogent and useful. I am indebted to Alan Tokunaga and to Johnathan Elias for their thoughtful responses to my invitation to participate. Their responses were read at the meeting and helped to inform the proceedings. Andy Young and Alan Clark have been especially helpful in supplying names, references, and ideas. Thanks must go also to Jeffrey Robbins, Physics Editor at Springer Verlag/Publishers for his help and support in getting these discussions a broader audience. Wendy Amero's typing helped to systematize the manuscripts. Any remaining errors of thought or text are entirely my responsibility.

References

Milone, E. F., Robb, R. M., Babott, F. M., and Hansen, C. H. 1982. Applied Optics, $\underline{21}$, 2992.

Milone, E.F., and Robb, R. M. 1983. Publs. Astron. Soc. Pacific, $\underline{95}$, 666.

Lecture Notes in Mathematics

Vol. 1236: Stochastic Partial Differential Equations and Applications. Proceedings, 1985. Edited by G. Da Prato and L. Tubaro. V, 257 pages. 1987.

Vol. 1237: Rational Approximation and its Applications in Mathematics and Physics. Proceedings, 1985. Edited by J. Gilewicz, M. Pindor and W. Siemaszko. XII, 350 pages. 1987.

Vol. 1250: Stochastic Processes – Mathematics and Physics II. Proceedings 1985. Edited by S. Albeverio, Ph. Blanchard and L. Streit. VI, 359 pages. 1987.

Vol. 1251: Differential Geometric Methods in Mathematical Physics. Proceedings, 1985. Edited by P. L. García and A. Pérez-Rendón. VII, 300 pages. 1987.

Vol. 1255: Differential Geometry and Differential Equations. Proceedings, 1985. Edited by C. Gu, M. Berger and R.L. Bryant. XII, 243 pages. 1987.

Vol. 1256: Pseudo-Differential Operators. Proceedings, 1986. Edited by H.O. Cordes, B. Gramsch and H. Widom. X, 479 pages. 1987.

Vol. 1258: J. Weidmann, Spectral Theory of Ordinary Differential Operators. VI, 303 pages. 1987.

Vol. 1260: N.H. Pavel, Nonlinear Evolution Operators and Semigroups. VI, 285 pages. 1987.

Vol. 1263: V.L. Hansen (Ed.), Differential Geometry. Proceedings, 1985. XI, 288 pages. 1987.

Vol. 1265: W. Van Assche, Asymptotics for Orthogonal Polynomials. VI, 201 pages. 1987.

Vol. 1267: J. Lindenstrauss, V.D. Milman (Eds.), Geometrical Aspects of Functional Analysis. Seminar. VII, 212 pages. 1987.

Vol. 1269: M. Shiota, Nash Manifolds. VI, 223 pages. 1987.

Vol. 1270: C. Carasso, P.-A. Raviart, D. Serre (Eds.), Nonlinear Hyperbolic Problems. Proceedings, 1986. XV, 341 pages. 1987.

Vol. 1272: M.S. Livšic, L.L. Waksman, Commuting Nonselfadjoint Operators in Hilbert Space. III, 115 pages. 1987.

Vol. 1273: G.-M. Greuel, G. Trautmann (Eds.), Singularities, Representation of Algebras, and Vector Bundles. Proceedings, 1985. XIV, 383 pages. 1987.

Vol. 1275: C.A. Berenstein (Ed.), Complex Analysis I. Proceedings, 1985–86. XV, 331 pages. 1987.

Vol. 1276: C.A. Berenstein (Ed.), Complex Analysis II. Proceedings, 1985–86. IX, 320 pages. 1987.

Vol. 1277: C.A. Berenstein (Ed.), Complex Analysis III. Proceedings, 1985–86. X, 350 pages. 1987.

Vol. 1283: S. Mardešić, J. Segal (Eds.), Geometric Topology and Shape Theory. Proceedings, 1986. V, 261 pages. 1987.

Vol. 1285: I.W. Knowles, Y. Saitō (Eds.), Differential Equations and Mathematical Physics. Proceedings, 1986. XVI, 499 pages. 1987.

Vol. 1287: E.B. Saff (Ed.), Approximation Theory, Tampa. Proceedings, 1985–1986. V, 228 pages. 1987.

Vol. 1288: Yu. L. Rodin, Generalized Analytic Functions on Riemann Surfaces. V, 128 pages. 1987.

Vol. 1294: M. Queffélec, Substitution Dynamical Systems – Spectral Analysis. XIII, 240 pages. 1987.

Vol. 1299: S. Watanabe, Yu. V. Prokhorov (Eds.), Probability Theory and Mathematical Statistics. Proceedings, 1986. VIII, 589 pages. 1988.

Vol. 1300: G.B. Seligman, Constructions of Lie Algebras and their Modules. VI, 190 pages. 1988.

Vol. 1302: M. Cwikel, J. Peetre, Y. Sagher, H. Wallin (Eds.), Function Spaces and Applications. Proceedings, 1986. VI, 445 pages. 1988.

Vol. 1303: L. Accardi, W. von Waldenfels (Eds.), Quantum Probability and Applications III. Proceedings, 1987. VI, 373 pages. 1988.

Lecture Notes in Physics

Vol. 317: C. Signorini, S. Skorka, P. Spolaore, A. Vitturi (Eds.), Heavy Ion Interactions Around the Coulomb Barrier. Proceedings, 1988. X, 329 pages. 1988.

Vol. 318: B. Mercier, An Introduction to the Numerical Analysis of Spectral Methods. V, 154 pages. 1989.

Vol. 319: L. Garrido (Ed.), Far from Equilibrium Phase Transitions. Proceedings, 1988. VIII, 340 pages. 1988.

Vol. 320: D. Coles (Ed.), Perspectives in Fluid Mechanics. Proceedings, 1985. VII, 207 pages. 1988.

Vol. 321: J. Pitowsky, Quantum Probability – Quantum Logic. IX, 209 pages. 1989.

Vol. 322: M. Schlichenmaier, An Introduction to Riemann Surfaces, Algebraic Curves and Moduli Spaces. XIII, 148 pages. 1989.

Vol. 323: D.L. Dwoyer, M.Y. Hussaini, R.G. Voigt (Eds.), 11th International Conference on Numerical Methods in Fluid Dynamics. XIII, 622 pages. 1989.

Vol. 324: P. Exner, P. Šeba (Eds.), Applications of Self-Adjoint Extensions in Quantum Physics. Proceedings, 1987. VIII, 273 pages. 1989.

Vol. 325: E. Brändas, N. Elander (Eds.), Resonances, Proceedings, 1987. XVIII, 564 pages. 1989.

Vol. 326: A. Grauel, Feldtheoretische Beschreibung der Thermodynamik für Grenzflächen. IX, 317 Seiten. 1989.

Vol. 327: K. Meisenheimer, H.-J. Röser (Eds.), Hot Spots in Extragalactic Radio Source. Proceedings, 1988, XII, 301 pages. 1989.

Vol. 328: G. Wegner (Ed.), White Dwarfs. Proceedings, 1988. XIV, 524 pages. 1989.

Vol. 329: A. Heck, F. Murtagh (Eds.), Knowledge Based Systems in Astronomy. IV, 280 pages. 1989.

Vol. 330: J.M. Moran, J.N. Hewitt, K.Y. Lo (Eds.), Gravitational Lenses. Proceedings, 1988. XIV, 238 pages. 1989.

Vol. 331: G. Winnewisser, J.T. Armstrong (Eds.), The Physics and Chemistry of Interstellar Molecular Clouds mm and Sub-mm Observations in Astrophysics. Proceedings, 1988. XVIII, 463 pages. 1989.

Vol. 332: P. Flin, H.W. Duerbeck (Eds.), Morphological Cosmology. Proceedings, 1988. VII, 438 pages. 1989.

Vol. 333: I. Appenzeller, H.J. Habing, P. Léna (Eds.), Evolution of Galaxies – Astronomical Observations. Proceedings, 1988. X, 391 pages. 1989.

Vol. 334: L. Maraschi, T. Maccacaro, M.-H. Ulrich (Eds.), BL Lac Objects. XIII, 497 pages. 1989.

Vol. 335: A. Lakhtakia, V.K. Varadan, V.V. Varadan, Time-Harmonic Electromagnetic Fields in Chiral Media. VII, 121 pages. 1989.

Vol. 336: M. Müller, Consistent Classical Supergravity Theories. VI, 125 pages. 1989.

Vol. 337: A.P. Maclin, T.L. Gill, W.W. Zachary (Eds.), Magnetic Phenomena. Proceedings, 1988. VI, 142 pages. 1989.

Vol. 338: V. Privman, N.M. Švrakić, Directed Models of Polymers, Interfaces, and Clusters: Scaling and Finite-Size Properties. VI, 120 pages. 1989.

Vol. 339: F. Ehlotzky (Ed.), Fundamentals of Laser Interactions II. Proceedings, 1989. XI, 317 pages. 1989.

Vol. 340: M. Peshkin, A. Tonomura, The Aharonov-Bohm Effect. VI, 152 pages. 1989.

Vol. 341: E.F. Milone (Ed.), Infrared Extinction and Standardization. Proceedings, 1988. III, 79 pages. 1989.